Selected Titles in This Series

597 **Ragnar-Olaf Buchweitz and John J. Millson,** CR-geometry and deformations of isolated singularities, 1997

596 **Paul S. Bourdon and Joel H. Shapiro,** Cyclic phenomena for composition operators, 1997

595 **Eldar Straume,** Compact connected Lie transformation groups on spheres with low cohomogeneity, II, 1997

594 **Solomon Friedberg and Hervé Jacquet,** The fundamental lemma for the Shalika subgroup of $GL(4)$, 1996

593 **Ajit Iqbal Singh,** Completely positive hypergroup actions, 1996

592 **P. Kirk and E. Klassen,** Analytic deformations of the spectrum of a family of Dirac operators on an odd-dimensional manifold with boundary, 1996

591 **Edward Cline, Brian Parshall, and Leonard Scott,** Stratifying endomorphism algebras, 1996

590 **Chris Jantzen,** Degenerate principal series for symplectic and odd-orthogonal groups, 1996

589 **James Damon,** Higher multiplicities and almost free divisors and complete intersections, 1996

588 **Dihua Jiang,** Degree 16 Standard L-function of $GSp(2) \times GSp(2)$, 1996

587 **Stéphane Jaffard and Yves Meyer,** Wavelet methods for pointwise regularity and local oscillations of functions, 1996

586 **Siegfried Echterhoff,** Crossed products with continuous trace, 1996

585 **Gilles Pisier,** The operator Hilbert space *OH*, complex interpolation and tensor norms, 1996

584 **Wayne W. Barrett, Charles R. Johnson, and Raphael Loewy,** The real positive definite completion problem: Cycle completability, 1996

583 **Jin Nakagawa,** Orders of a quartic field, 1996

582 **Darryl McCollough and Andy Miller,** Symmetric automorphisms of free products, 1996

581 **Martin U. Schmidt,** Integrable systems and Riemann surfaces of infinite genus, 1996

580 **Martin W. Liebeck and Gary M. Seitz,** Reductive subgroups of exceptional algebraic groups, 1996

579 **Samuel Kaplan,** Lebesgue theory in the bidual of $C(X)$, 1996

578 **Ale Jan Homburg,** Global aspects of homoclinic bifurcations of vector fields, 1996

577 **Freddy Dumortier and Robert Roussarie,** Canard cycles and center manifolds, 1996

576 **Grahame Bennett,** Factorizing the classical inequalities, 1996

575 **Dieter Heppel, Idun Reiten, and Sverre O. Smalø,** Tilting in Abelian categories and quasitilted algebras, 1996

574 **Michael Field,** Symmetry breaking for compact Lie groups, 1996

573 **Wayne Aitken,** An arithmetic Riemann-Roch theorem for singular arithmetic surfaces, 1996

572 **Ole H. Hald and Joyce R. McLaughlin,** Inverse nodal problems: Finding the potential from nodal lines, 1996

571 **Henry L. Kurland,** Intersection pairings on Conley indices, 1996

570 **Bernold Fiedler and Jürgen Scheurle,** Discretization of homoclinic orbits, rapid forcing and "invisible" chaos, 1996

569 **Eldar Straume,** Compact connected Lie transformation groups on spheres with low cohomogeneity, I, 1996

568 **Raúl E. Curto and Lawrence A. Fialkow,** Solution of the truncated complex moment problem for flat data, 1996

567 **Ran Levi,** On finite groups and homotopy theory, 1995

(*Continued in the back of this publication*)

CR-Geometry and Deformations of Isolated Singularities

MEMOIRS
of the
American Mathematical Society

Number 597

CR-Geometry and Deformations of Isolated Singularities

Ragnar-Olaf Buchweitz
John J. Millson

January 1997 • Volume 125 • Number 597 (third of 5 numbers) • ISSN 0065-9266

American Mathematical Society
Providence, Rhode Island

1991 *Mathematics Subject Classification.*
Primary 32S30, 32F40, 14B12.

Library of Congress Cataloging-in-Publication Data

Buchweitz, Ragnar-Olaf, 1952–
CR-geometry and deformations of isolated singularities / Ragnar-Olaf Buchweitz, John J. Millson.
p. cm.—(Memoirs of the American Mathematical Society, ISSN 0065-9266 ; no. 597)
"January 1997, volume 125, number 597 (third of 5 numbers)."
Includes bibliographical references.
ISBN 0-8218-0541-X (alk. paper)
1. CR submanifolds. 2. Singularities (Mathematics) I. Millson, John J. (John James), 1946– . II. Title. III. Series.
QA3.A57 no. 597
[QA649]
510 s—dc20
[516.3′6] 96-44758
CIP

Memoirs of the American Mathematical Society

This journal is devoted entirely to research in pure and applied mathematics.

Subscription information. The 1997 subscription begins with number 595 and consists of six mailings, each containing one or more numbers. Subscription prices for 1997 are $414 list, $331 institutional member. A late charge of 10% of the subscription price will be imposed on orders received from nonmembers after January 1 of the subscription year. Subscribers outside the United States and India must pay a postage surcharge of $30; subscribers in India must pay a postage surcharge of $43. Expedited delivery to destinations in North America $35; elsewhere $110. Each number may be ordered separately; *please specify number* when ordering an individual number. For prices and titles of recently released numbers, see the New Publications sections of the *Notices of the American Mathematical Society.*

Back number information. For back issues see the *AMS Catalog of Publications.*

Subscriptions and orders should be addressed to the American Mathematical Society, P. O. Box 5904, Boston, MA 02206-5904. *All orders must be accompanied by payment.* Other correspondence should be addressed to Box 6248, Providence, RI 02940-6248.

Memoirs of the American Mathematical Society is published bimonthly (each volume consisting usually of more than one number) by the American Mathematical Society at 201 Charles Street, Providence, RI 02904-2294. Periodicals postage paid at Providence, RI. Postmaster: Send address changes to Memoirs, American Mathematical Society, P. O. Box 6248, Providence, RI 02940-6248.

This publication is indexed in *Science Citation Index*®, *SciSearch*®, *Research Alert*®, *CompuMath Citation Index*®, *Current Contents*®/*Physical, Chemical & Earth Sciences.*
Printed in the United States of America.

♾ The paper used in this book is acid-free and falls within the guidelines established to ensure permanence and durability.

10 9 8 7 6 5 4 3 2 1 01 00 99 98 97

Contents

ABSTRACT. In this paper we show how to compute the parameter space X for the versal deformation of an isolated singularity $(V,0)$ — whose existence was shown by Grauert in 1972, — under the assumptions $\dim V \geq 4$, $\operatorname{depth}_{\{0\}} V \geq 3$, from the CR-structure on a link M of the singularity. We do this by showing that the space X is isomorphic to the space (denoted here by $\mathcal{K}_M$) associated to M by Kuranishi in 1977. In fact we produce isomorphisms of the associated complete local rings by producing quasi-isomorphisms of the controlling differential graded Lie algebras for the corresponding formal deformation theories.

The desired quasi-isomorphism corresponds to the diagram of deformation theories

$$\mathrm{Def}(V,0) \longleftarrow \mathrm{Def}(V) \longrightarrow \mathrm{Def}(U) \longrightarrow \mathrm{Def}(M)$$

where $U = V - \{0\}$.

Date: Received by the editor October 11, 1993.

1991 *Mathematics Subject Classification.* 32S30, 32F40, 14B12.

Key words and phrases. CR-geometry, Flat deformations of isolated singularities, Controlling differential graded Lie algebras, one-quasi-isomorphisms, Kodaira-Spencer algebra, Tangent complex, Kuranishi's CR-deformation theory.

The first author was supported by an NSERC grant 3-642-114-80; the second by National Science Foundation grant DMS-90-02116.

0. Introduction

In this paper we show how to compute the analytic parameter space for the versal deformation of an isolated singularity $(V, 0)$ — whose existence was shown by H. Grauert, [Gr1], in 1972, from the CR-structure on a link M of the singularity. There are two ideas developed in this paper. The first was provided by the fundamental work [K2] of Kuranishi in 1977. In this paper Kuranishi wrote down a first order non-linear system of partial differential equations on a link of the singularity with a finite dimensional formal solution space assuming $\dim V \geq 3$ (see Theorem 3.2 for a precise statement). This space had the key property that it did not depend on the choice of sphere used to define the link. In 1983, in [Mi1], Miyajima proved that Kuranishi's space had a complex analytic structure (provided $\dim V \geq 4$) to be denoted $\mathcal{K}_M$ henceforth. His work was based on earlier work of Akahori [Ak1]–[Ak4]. Our main theorem below relating $\mathcal{K}_M$ to the parameter space of the versal deformation of $(V, 0)$ was obtained independently by Miyajima [Mi2]. His result was based in part on [Fu]. Roughly, [Fu] is concerned with passing from the deformation theory of $(V, 0)$ to that of the complement of 0 in a Stein representative for $(V, 0)$ (Chapters 2, 4, 5, 6, 7 of this work) and [Mi2] with passing from the deformation theory of the complement to that of the link (Chapters 3, 8 and 9 of this work). Unfortunately [Fu] is not available at present. Fujiki and Miyajima also obtain results relating the versal families.

The second idea is the principle that deformation problems should be controlled by differential graded Lie algebras which we learned from Deligne [D] — similar ideas are to be found in [SS1]. According to this principle, in order to prove that two deformation spaces are isomorphic one finds controlling differential graded Lie algebras (see Chapter 1) and then proves these algebras are "one-quasi-isomorphic", see [GM1] or [M1] for other applications of this principle. Here two differential graded Lie algebras L and $\overline{L}$ are defined to be one-quasi- isomorphic if there exists a chain of homomorphisms $L = L_1 \longrightarrow L_2 \longleftarrow L_3 \longrightarrow \cdots \longrightarrow L_n = \overline{L}$ such that each arrow induces an isomorphism on H^1 and an injection on H^2. This is what we do in this paper. Our main theorem goes as follows.

THEOREM. *Suppose $(V, 0)$ is a normal isolated singularity and satisfies*

(1) $\dim V \geq 4$,
(2) $\mathrm{depth}_{\{0\}} V \geq 3$.

Then the base space of the versal deformation of $(V, 0)$ is isomorphic to $\mathcal{K}_M$.

Remarks. The assumption (2) is equivalent to the assumption that $H^1(U, \mathcal{O}) = \{0\}$ where $U = V - \{0\}$. If we do not assume (2), we show that the base space for the versal deformation of $(V, 0)$ is isomorphic to a closed subgerm of $\mathcal{K}_M$. In chapter 10 we give examples such that $(V, 0)$ is normal but the deformation space of $(V, 0)$ is a proper subgerm of $\mathcal{K}_M$.

Our proof involves many technical details but the basic motivation is simple. It is to prove that the arrows in the following diagram of *formal* deformation theories

$$\mathrm{Def}(V,0) \longleftarrow \mathrm{Def}(V) \longrightarrow \mathrm{Def}(U) \longrightarrow \mathrm{Def}(M)$$

are isomorphisms. The middle arrow is the restriction map of Schlessinger, [Sc2] or [Ar2], Part I, §9. A formal deformation of V is a sheaf S on V satisfying certain axioms. Then the middle arrow restricts S to U. We remind the reader that by Artin's Theorem, [Ar1, pg. 282], formal isomorphism of analytic germs implies analytic isomorphism.

In fact we study the corresponding diagram of maps of controlling differential graded Lie algebras and prove they are all one-quasi-isomorphisms. A critical intermediate step is Theorem D below which allows us to replace the tangent complex $L_U^{\bullet}$ of U by the Kodaira-Spencer algebra $\mathcal{L}^{\bullet}(U) = \mathcal{A}^{0,\bullet}(U, T^{1,0}(U))$ which is needed to compare with Kuranishi's theory.

Our proof divides into six theorems, some (especially Theorems C and D) of interest in their own right. In what follows $L_{V,0}$ denotes the tangent complex of the germ $(V,0)$ and L_V (resp. L_U) the tangent complex of the complex analytic space V (resp. U). Also $\overline{\mathcal{K}}$ is the differential graded Lie algebra on M constructed in Chapter 9. Our six theorems are then the following.

THEOREM A (Chapter 5). *$L_{V,0}$ controls the deformation theory of $(V,0)$.*

THEOREM B (Chapter 5). *L_V is one-quasi-isomorphic to $L_{V,0}$.*

THEOREM C (Chapter 6). *Let V be a complex analytic space. Then L_V controls the deformation theory of V.*

COROLLARY [SC2]. *The rings $R_{L_{V,0}}$ and R_{L_U} (see Chapter 1) are isomorphic.*

THEOREM D (Chapter 7). *If U is a complex manifold, L_U and the Kodaira-Spencer algebra $\mathcal{A}^{0,\bullet}(U, T^{1,0}(U))$ are quasi-isomorphic.*

COROLLARY. *The Kodaira-Spencer algebra controls the (formal) deformation theory of U (even if U is not compact).*

THEOREM E (Chapter 9). *The Kodaira-Spencer algebra of U and $\overline{\mathcal{K}}$ are one-quasi-isomorphic.*

THEOREM F (Chapter 9). *$\overline{\mathcal{K}}$ controls Kuranishi's CR-deformation theory.*

An outline of this paper without detailed proofs was published in [M3].

ACKNOWLEDGEMENTS

This paper could not have been written without many conversations with other mathematicians; among them, Bill Goldman, Steve Halperin and Mike Schlessinger. We would especially like to thank Madhav Nori for providing us with the material concerning infinite dimensional affine varieties in Chapter 1, Jack Lee for many conversations about CR-manifolds and H. Flenner for help with the counterexamples in Chapter 10. We have made extensive use of Flenner's thesis [F] in this paper. Finally the second author would like to thank Pierre Deligne for outlining the connection between differential graded Lie algebras and deformation theory in [D] some years ago.

NOTATIONAL CONVENTIONS

Throughout this paper $\boldsymbol{k}$ will denote a field of characteristic zero, usually the complex numbers. By an Artin local $\boldsymbol{k}$–algebra will mean a local $\boldsymbol{k}$–algebra A that is finite dimensional as a $\boldsymbol{k}$–vector space and such that the residue field of A is $\boldsymbol{k}$. The set of such algebras comprise the objects in a category to be denoted $\mathcal{A}$.

In Chapters 4 and 6 we have used a subscript $_*$ to indicate a simplicial object relative to the nerve $\mathcal{N}$ of a fixed covering. For example a simplicial complex space X_* is a contravariant functor from $\mathcal{N}$ considered as a category in the usual way to the category of complex spaces. Thus for each simplex $\alpha \in N$ we are given a complex space X_α and for each inclusion $\alpha \subset \beta$ we are given a morphism $p_{\alpha\beta} : X_\beta \longrightarrow X_\alpha$ satisfying $p_{\alpha\alpha} = id$ and $p_{\alpha\beta} \circ p_{\beta\gamma} = p_{\alpha\gamma}$ for $\alpha \subset \beta \subset \gamma$. The reader should note the non-standard definition in Chapter 4 (taken from [F], page 33) of a *free* module over the structure sheaf $\mathcal{O}_{W_*}$ of a simplicial ringed space $(W_*, \mathcal{O}_{W_*})$.

All manifolds will be assumed to be C^∞, connected and paracompact and all tensor fields will be assumed to be C^∞ unless the contrary is stated explicitly. We will use the notation $\mathcal{A}^{\bullet}(M)$ for the complex de Rham algebra of a manifold M and $\mathcal{A}^{\bullet}$ for the corresponding sheaf. We will adopt similar notation for the Dolbeault algebra $\mathcal{A}^{0,\bullet}(M)$ and the corresponding sheaf $\mathcal{A}^{0,\bullet}$ for a complex manifold M. Finally we will use the notation $\mathcal{L}^{\bullet}(M)$ for the Kodaira-Spencer algebra $\mathcal{A}^{0,\bullet}\left(M, T^{1,0}(M)\right)$ of a complex manifold M. We will use $\mathcal{L}^{\bullet}$ for the corresponding sheaf until Chapter 8 where $\mathcal{L}^{\bullet}$ will be used to abbreviate $\mathcal{L}^{\bullet}(M)$.

If $V^{\bullet}$ is a graded vector space and $m \in \mathbb{Z}$ then $V^{\bullet}[m]$ will denote the graded vector space such that $(V[m])^n = V^{n+m}$. If V is an (ungraded) vector space, we will often identify it with the graded vector space with zero graded piece equal to V and all other graded pieces equal to zero. By $S(V)$ we denote the free graded commutative algebra generated by the graded vector space V. In this paper we will have occasion to consider infinite direct sums $\bigoplus_{i=0}^{\infty} E_i$ of graded vector bundles (and sheaves) such that the degree of E_i is i, over a manifold M. We will define the global sections $\Gamma\left(M, \bigoplus_{i=0}^{\infty} E_i\right)$ of such a bundle (or sheaf) by

$$\Gamma\left(M, \bigoplus_{i=0}^{\infty} E_i\right) = \bigoplus_{i=0}^{\infty} \Gamma(M, E_i).$$

This is consistent with the analogous definition for (graded) Hom in Chapter 1.

1. Controlling Differential Graded Lie Algebras

In this chapter we study the deformation theory associated to a differential graded Lie algebra over a field $\boldsymbol{k}$ of characteristic zero, see [GM1] for details. We begin with some definitions.

A *graded Lie algebra* over $\boldsymbol{k}$ will mean a $\boldsymbol{k}$–vector space

$$L = \bigoplus_i L^i$$

graded by the integers satisfying (graded) skew-commutativity

$$[\alpha, \beta] + (-1)^{ij} [\beta, \alpha] = 0$$

and the graded Jacobi identity

$$(-1)^{ik} [\alpha, [\beta, \gamma]] + (-1)^{ji} [\beta, [\gamma, \alpha]] + (-1)^{kj} [\gamma, [\alpha, \beta]] = 0$$

where $\alpha \in L^i$, $\beta \in L^j$, $\gamma \in L^k$.

A *(graded) derivation of degree ℓ* consists of a family of linear maps $d : L^i \longrightarrow L^{i+\ell}$ satisfying

$$d[\alpha, \beta] = [d\alpha, \beta] + (-1)^{i\ell}[\alpha, d\beta]$$

where $\alpha \in L^i$, $\beta \in L$.

A *differential graded Lie algebra* is pair (L, d) where L is a graded Lie algebra and d is a derivation of degree 1 such that the composition $d \circ d = 0$. Thus the cohomology algebra $H^{\cdot}(L) = \bigoplus_{i \geq 0} H^i(L)$ is defined and inherits the structure of a graded Lie algebra. A large number of the differential graded Lie algebras occurring in this paper will arise as follows.

Let $V = \bigoplus_{i \geq 0} V^i$ be a graded vector space over $\boldsymbol{k}$. An endomorphism T of V of degree ℓ will be a family $T : V^i \longrightarrow V^{i+\ell}$. The space of all such T will be denoted $\mathrm{Hom}^{\ell}(V, V)$. Throughout this paper the symbol $\mathrm{Hom}(V, V)$, where V is as above, will denote the *direct sum*

$$\mathrm{Hom}(V, V) = \bigoplus_{\ell} \mathrm{Hom}^{\ell}(V, V).$$

Then $\mathrm{Hom}(V, V)$ is a graded Lie algebra under the rule

$$[S, T] = S \circ T - (-1)^{ij} T \circ S$$

where $S \in \mathrm{Hom}^i(V, V)$, $T \in \mathrm{Hom}^j(V, V)$. In the event V is a graded (commutative) algebra [GrMo], page 109, we restrict to those homomorphisms that are graded derivations. This space will be denoted $\mathrm{Der}\, V$. It is immediate that $\mathrm{Der}\, V$ is closed under $[\ ,\ ]$. The above spaces are graded by the integers, we will usually restrict to those of non-negative degree to be denoted $\mathrm{Der}^+ V$.

One of the main points of [GM1] is that there is a deformation functor $\mathcal{C}(L; \cdot)$ (precisely a groupoid cofibered over $\mathcal{A}$) canonically associated to any differential graded Lie algebra. We describe $\mathcal{C}(L; \cdot)$ as follows. Let $A \in \mathrm{Obj}\ \mathcal{A}$ with maximal ideal $\mathfrak{m}$. We define the groupoid $\mathcal{C}(L;A)$ as follows.

$$\text{Obj } \mathcal{C}(L;A) = \left\{\eta \in L^1 \otimes \mathfrak{m} : d\eta + \tfrac{1}{2}[\eta,\eta] = 0\right\}$$

$$\text{Mor } \mathcal{C}(L;A) = \exp(L^0 \otimes \mathfrak{m})$$

where $\exp(L^0 \otimes \mathfrak{m})$ is the nilpotent Lie group with underlying space $L^0 \otimes \mathfrak{m}$ and equipped with the Campbell-Baker-Hausdorff multiplication

$$(X,Y) \longrightarrow \log\left(\exp(X) \cdot \exp(Y)\right).$$

The morphisms act on the objects by the "affine action" [GM1], (1.3). The affine action will be denoted $\alpha(e^\lambda) \cdot \eta$ for $\lambda \in L^0 \otimes \mathfrak{m}$, $\eta \in L^1 \otimes \mathfrak{m}$. The action α is determined by the formula

$$d\alpha(\lambda) \cdot \eta = [\lambda, \eta] - d\lambda.$$

We let Iso $\mathcal{C}(L;A)$ denote the set of isomorphism classes. We observe that if $\phi : L \longrightarrow \overline{L}$ is a homomorphism of differential graded Lie algebras then ϕ induces natural transformations of functors from $\mathcal{A}$ to **Sets**

$$\begin{aligned} \phi &: \text{Obj } \mathcal{C}(L;\cdot) \longrightarrow \text{Obj } \mathcal{C}(\overline{L};\cdot) \\ \phi &: \text{Iso } \mathcal{C}(L;\cdot) \longrightarrow \text{Iso } \mathcal{C}(\overline{L};\cdot). \end{aligned}$$

The functor Iso $\mathcal{C}(L;\cdot)$ is not pro-representable in general but it does have a hull if $\dim H^1(L) < \infty$. We recall what this means. Let F and G be functors on $\mathcal{A}$ with values in **Sets** and let $\eta : F \longrightarrow G$ be a natural transformation. Then η is smooth if for any surjection $A \longrightarrow \overline{A}$ in $\mathcal{A}$ the induced map

$$F(A) \longrightarrow F(\overline{A}) \otimes_{G(\overline{A})} G(A)$$

is surjective. By the principle of Artinian induction, [GM1], 2.5, it suffices to check this in the special case $\overline{A} = A/\mathfrak{J}$ where $\mathfrak{J} \subset A$ is an ideal satisfying $\mathfrak{J}m = 0$. A smooth natural transformation η is minimally smooth if given $A \in \text{Obj } \mathcal{A}$ with maximal ideal $\mathfrak{m}$ satisfying $\mathfrak{m}^2 = 0$ then $\eta_A : F(A) \longrightarrow G(A)$ is an isomorphism.

Now let $\mathcal{C}$ be the category of complete local Noetherian $\mathbf{k}$–algebras. Let $R \in \mathcal{C}$ and define $h_R : \mathcal{C} \longrightarrow \textbf{Sets}$ by

$$h_R(S) = \text{Hom}_{\mathbf{k}-\text{alg}}(R,S).$$

Now suppose F is as above and extend F to $\mathcal{C}$ by the formula

$$F(R) = \varprojlim_n F(R/\mathfrak{m}^n)$$

where $\mathfrak{m} \subset R$ is the maximal ideal. We observe that if $u \in \text{Hom}_{\mathbf{k}-\text{alg}}(R,A)$ there is an induced map $F(u) : F(R) \longrightarrow F(A)$. Hence given $\xi \in F(R)$ we obtain a natural transformation $e_\xi : h_R \longrightarrow F$ of functors on $\mathcal{A}$ given by

$$e_\xi(u) = F(u)(\xi).$$

We have two important definitions from [Sc1]. A complete local $\mathbf{k}$–algebra R together with an element $\xi \in F(R)$ is said to *pro-represent* F if the natural transformation e_ξ is an isomorphism of functors.

A complete local $\mathbf{k}$–algebra R together with an element $\xi \in F(R)$ is said to be a *hull* of F if the natural transformation $e_\xi : h_R \longrightarrow F$ is minimally smooth.

We have the following result, Proposition 2.8 of [Sc1].

Lemma 1.1. *Let (R,ξ) and (R',ξ') be hulls of F. Then there exists an isomorphism $u : R \longrightarrow R'$ such that $F(u)(\xi) = \xi'$.*

We can now state the main theorem of this section.

Theorem 1.1. *Suppose L is a differential graded Lie algebra satisfying*

$$\dim H^1(L) < \infty.$$

Then Iso $\mathcal{C}(L;\cdot)$ *has a hull* (R_L,ξ).

This theorem may be proved by applying Theorem 2.10 of [Sc1]. However we will give a direct proof which we owe to M. Nori. The ideas which we need to give this proof will be useful in Chapter 9.

Let L be a differential graded Lie algebra with $\dim H^1(L) < \infty$. Choose a complement C^1 in L^1 to the 1-coboundaries $B^1 = dL^0$. We construct a functor $Y_L : \mathcal{A} \longrightarrow$ **Sets** as follows. Let $A \in \text{Obj } \mathcal{A}$ and let $\mathfrak{m} \subset A$ be the maximal ideal of A. Then we define $Y_L(A)$ by

$$Y_L(A) = \left\{\eta \in C^1 \otimes \mathfrak{m} : d\eta + \tfrac{1}{2}[\eta,\eta] = 0\right\}.$$

Remark. The definition of Y_L is motivated by Kuranishi's construction of the versal deformation space of a complex manifold.

Our initial goal is to prove that the functor Y_L is pro-representable. In order to prove this and in order to compare Y_L with the analytic constructions of Kuranishi and Miyajima in Chapter 10 we will need a careful treatment of A–points in infinite dimensions. The following results relating A–points and the ring A_L were shown to us by Madhav Nori.

We begin by defining a map $\varphi : L \longrightarrow M$ of vector spaces to be homogeneous of degree k if there is a k–linear map $b : L \times \cdots \times L \longrightarrow M$ such that $\varphi(v) = b(v, v, \ldots, v)$. In what follows the notation $L' < L$ will mean that L' is a finite-dimensional subspace of L. We define $\text{Pol}(L, M)$, the space of polynomial mappings from one (possibly infinite-dimensional) vector space to another by

$$\text{Pol}(L,M) = \varprojlim_{L'<L} \text{Pol}(L',M).$$

As usual $\varphi \in \text{Pol}(L',M)$ for $\dim L' < \infty$ means that φ is a finite sum of homogeneous maps. Thus $\text{Pol}(L,M)$ consists of those maps $\varphi : L \longrightarrow M$ whose restriction to every finite dimensional subspace of L is polynomial. We define $\deg\varphi$ for $\varphi \in \text{Pol}(L,M)$ to be the least upper bound of the numbers $\{\deg(\varphi|L') : L' < L\}$. Thus $\deg\varphi$ could be ∞. The polynomial maps of finite degree are those obtained by taking a finite sum of homogeneous maps. We note that the map $\varphi : L^1 \longrightarrow L^2$ above given by $\varphi(\eta) = d\eta + \frac{1}{2}[\eta,\eta]$ is a polynomial map of degree 2.

We define the "affine ring" A_L of L by

$$A_L = \text{Pol}(L, \boldsymbol{k}).$$

We give $\mathrm{Pol}(L', \boldsymbol{k})$ the discrete topology, for $L' < L$, whence A_L becomes a topological $\boldsymbol{k}$–algebra. Now let A be any $\boldsymbol{k}$–algebra, $f \in A_L$ and $\tau \in L \otimes A$. We define $f(\tau) \in A$ as follows. First there exists $L' < L$ such that $\tau \in L' \otimes a$. Now f induces an element of $\mathrm{Sym}((L')^*)$. Thus it suffices to define $f_k(\eta \otimes x)$ where f_k is homogeneous of degree k, $\eta \in L'$ and $x \in A$.

We define

$$f_k(\eta \otimes x) = f_k(\eta) x^k.$$

Let $\mathrm{CAH}(A_L, A)$ denote the space of continuous algebra homomorphisms where A is given the discrete topology. It is immediate that e_τ defined by $e_\tau(f) = f(\tau)$ is an element of $\mathrm{CAH}(A_L, A)$.

Lemma 1.2. *The map* $e : L \otimes A \longrightarrow \mathrm{CAH}(A_L, A)$ *defined by* $e(\tau) = e_\tau$ *is a bijection.*

Proof. Since A is discrete the natural map

$$\varinjlim_{L'<L} \mathrm{CAH}(A_L, A) \longrightarrow \mathrm{CAH}\left(\varprojlim_{L'<L} A_{L'}, A\right)$$

is a bijection and the lemma follows because e is the limit of the bijections

$$L' \otimes A \longrightarrow \mathrm{CAH}(A_{L'}, A)$$

over $L' < L$. □

Now suppose $\varphi : L \longrightarrow M$ is a polynomial map. We construct a map $\tilde{\varphi} : A_M \longrightarrow A_L$ as follows. Let $L' < L$. Observe that $\varphi(L')$ is contained in a finite dimensional subspace M' of M. Let $f \in A_M$. We then define the image of $\tilde{\varphi}(f)$ in $A_{L'}$ to be $f \circ \varphi | L'$. The proof of the next lemma is similar to that of Lemma 1.2 and is left to the reader.

Lemma 1.3. *The map* $\varphi \longrightarrow \tilde{\varphi}$ *induces an isomorphism*

$$\mathrm{Pol}(L, M) \longrightarrow \mathrm{CAH}(A_M, A_L).$$

If A is a $\boldsymbol{k}$–algebra, we obtain an induced polynomial map $\varphi_A : L \otimes A \longrightarrow M \otimes A$ given by

$$\varphi_A(\tau) = e_\tau \circ \tilde{\varphi} \quad \text{for } \tau \in L \otimes A.$$

The following lemma is useful for calculating φ_A directly. We again leave the proof to the reader.

Lemma 1.4. *Suppose* $\varphi : L \longrightarrow M$ *is homogeneous of degree* k, $\eta \in L$ *and* $x \in A$. *Then*

$$\varphi_A(\eta \otimes x) = \varphi(\eta) \otimes x^k.$$

We next want to discuss the affine ring of an affine subvariety. Let $\mathfrak{m}_M$ denote the (maximal) ideal in A_M consisting of functions vanishing at $0 \in M$. Since $\boldsymbol{k}$ has the discrete topology and e_0 is continuous, the ideal $\mathfrak{m}_M$ is open and closed. Let $\varphi : L \longrightarrow M$ be a polynomial map and consider the "affine variety" $Y \subset L$ defined by $Y = \{\eta \in L : \varphi(\eta) = 0\}$. We have an associated functor Y on the category of $\boldsymbol{k}$–algebras defined by

$$Y(A) = \{\tau \in L \otimes A : \varphi_A(\tau) = 0\}.$$

Now let $I \subset A_L$ be the closure of the ideal generated by $\tilde{\varphi}(f)$, $f \in \mathfrak{m}_M$.

Lemma 1.5. *The evaluation map* $e : L \otimes A \longrightarrow \mathrm{CAH}(A_L, A)$ *induces a bijection* $e : Y(A) \longrightarrow \mathrm{CAH}(A_L/I, A)$.

Proof. We note $e_\alpha \circ \varphi = e_{\varphi_A(\alpha)}$. Then e_α annihilates $I \iff e_\alpha \circ \tilde{\varphi}$ annihilates $\mathfrak{m}_M \iff e_\alpha \circ \tilde{\varphi} = e_0 \iff e_{\varphi_A(\alpha)} = e_0$. □

Finally we discuss A–points of completions for $a \in \mathrm{Obj}\ \mathcal{A}$ (or $\mathcal{C}$). Clearly $\bigcap_{n=1}^{\infty} \mathfrak{m}_L^n = \{0\}$ and the completion $\hat{A}_L$ of A_L at $\mathfrak{m}_L$ is Hausdorff. If $L' < L$ we have a surjection $A_L/\mathfrak{m}_L^n \longrightarrow A_{L'}/(\mathfrak{m}_{L'})^n$ and an induced isomorphism of topological algebras $\hat{A}_L = \varprojlim_{L'<L} \hat{A}_{L'}$, where $\hat{A}_{L'}$ has the $\mathfrak{m}_{L'}$–adic topology. Now let $A \in \mathrm{Obj}\ \mathcal{C}$ with maximal ideal $\mathfrak{m}_A$. Since $A_L \subset \hat{A}_L$ is dense we have an embedding $\iota : \mathrm{CAH}(\hat{A}_L, A) \longrightarrow \mathrm{CAH}(A_L, A)$ induced by restriction. Let $\pi : A \longrightarrow \boldsymbol{k}$ be reduction modulo $\mathfrak{m}_A$.

Lemma 1.6. *The following diagram is commutative*

$$\begin{array}{ccc} L \otimes \mathfrak{m}_A & \longrightarrow & L \otimes A \\ \downarrow & & \downarrow \\ \mathrm{CAH}(\hat{A}_L, A) & \longrightarrow & \mathrm{CAH}(A_L, A)\ . \end{array}$$

Proof. We first observe that we have a commutative diagram

$$\begin{array}{ccc} L \otimes A & \longrightarrow & \mathrm{CAH}(A_L, A) \\ \downarrow & & \downarrow \\ L \otimes \boldsymbol{k} & \longrightarrow & \mathrm{CAH}(A_L, \boldsymbol{k}) \end{array}$$

where the vertical arrows are the maps induced by π and the horizontal arrows are the evaluation maps. Indeed the diagram is equivalent to the formula $(\pi \circ f)(\tau) = f\big(\pi(\tau)\big)$ for $\tau \in L \otimes A$, $f \in A_L$. This latter formula is immediate by taking $\tau = \eta \otimes x$, $\eta \in L$, $x \in A$. As a consequence of this diagram we see that under the evaluation map $L \otimes \mathfrak{m}_A$ corresponds to $\chi \in \mathrm{CAH}(A_L, A)$ such that $\pi \circ \chi = e_0$.

Now let $\chi \in \mathrm{CAH}(A_L, A)$. We claim that χ extends continuously to $\hat{A}_L \iff \chi(\mathfrak{m}_L) \subset \mathfrak{m}_A \iff \pi \circ \chi = e_0$. Sufficiency is clear. For necessity observe that if χ is continuous then $\pi \circ \chi$ factors through $\hat{A}_L/\mathfrak{m}_L^n$ for some n (since $\boldsymbol{k}$ has the discrete topology). But the image of $\mathfrak{m}_L$ in this quotient is nilpotent and 0 is the only nilpotent in $\boldsymbol{k}$. Hence $\pi \circ \chi(\mathfrak{m}_L) = 0$. □

In Chapter 9 we will need the space of formal maps $\mathrm{For}(L, M)$ defined by

$$\mathrm{For}(L, M) = \varprojlim_{L'<L} \mathrm{For}(L', M).$$

Thus $\varphi \in \mathrm{For}(L, M)$ means φ is a sum $\varphi = \sum_{k=1}^{\infty} \varphi_k$ where φ_k is homogeneous of degree k. The reader will verify that $\mathrm{For}(L, k)$ is isomorphic to $\hat{A}_L$. Of course a formal map φ does not induce a map from L to M but it does induce a map

$$\varphi_A : \mathrm{CAH}(\hat{A}_L, A) = L \otimes \mathfrak{m}_A \longrightarrow \mathrm{CAH}(\hat{A}_L, A) = M \otimes \mathfrak{m}_A$$

for $A \in \mathrm{Obj}\ \mathcal{A}$, which can be calculated using the formula in Lemma 1.4. We note that if $\mathfrak{m}_A^{n+1} = 0$ the map φ_A above is induced by the polynomial map $j^n\varphi$ of degree n given by $j^n\varphi = \sum_{k=1}^{n} \varphi_k$. Thus it is apparent that $(\varphi \circ \psi)_A = \varphi_A \circ \psi_A$ for $\psi \in \mathrm{For}(L, M)$, $\varphi \in \mathrm{For}(M, N)$, $A \in \mathrm{Obj}\ \mathcal{A}$.

Remark. In this paper formal maps will arise as follows. Suppose L and M are Banach spaces, B and B' are balls around the origin in L and M respectively and $\varphi : B \longrightarrow B'$ is an analytic map with $\varphi(0) = 0$. This means φ has a representation $\varphi = \sum_{k=1}^{\infty} \varphi_k$ with φ_k homogeneous of degree k and $\sum_{k=1}^{\infty} \varphi_k(x)$ normally convergent for $x \in B$. Clearly $\sum_{k=1}^{\infty} \varphi_k \in \mathrm{For}(L, M)$.

We now prove Y_L above is pro-representable as a consequence of the following more general theorem which we owe to M. Nori.

Theorem 1.2. *Suppose $\varphi : L \longrightarrow M$ is a polynomial map with $\varphi(0) = 0$ and $\ker d\varphi(0)$ is finite-dimensional. Then the functor $Y : \mathcal{A} \longrightarrow$* **Sets** *defined by*

$$Y(A) = \{\tau \in L \otimes \mathfrak{m}_A : \varphi_A(\tau) = 0\}$$

is pro-representable.

Proof. Let I be the closure of $\tilde{\varphi}(\mathfrak{m}_M)$ in $\hat{A}_L$ and let $R_L = \hat{A}_L / I$. By Lemmas 1.5 and 1.6 we see that $e : Y \longrightarrow \mathrm{CAH}(R_L, \cdot)$ is an isomorphism of functors. The local algebra R_L is obviously Hausdorff and complete. Let $\mathfrak{m}$ be its maximal ideal. We claim $\dim_k \mathfrak{m}/\mathfrak{m}^2 < \infty$. Indeed let $H = \ker d\varphi(0)$ and choose a complement K to H in L. Let $\alpha_1, \alpha_2, \ldots \alpha_n$ be the lift of a basis for H^* to L^*. Let $p : L \longrightarrow K$ be the projection. Then any $f \in \hat{A}_L$ has a unique Taylor series representation [Bo2], Ch. IV, §4, No. 5,

$$f(x) = \sum_{\nu} (D^\nu f)(p(x)) \alpha_1(x)^{\nu_1} \cdots \alpha_n(x)^{\nu_n}$$

and consequently the ideal of formal power series vanishing on K is the ideal $(\alpha_1, \alpha_2, \ldots, \alpha_n)$ in $\hat{A}_L$ generated by $\{\alpha_1, \alpha_2, \ldots, \alpha_n\}$. Put $T = d\varphi(0)$ whence $T|K$ is injective and consequently $S^i(T|K) : S^iK \longrightarrow S^iM$ is injective for all i. Hence the dual map $(S^iM)^* \longrightarrow (S^iK)^*$ is onto for all i. Now let $f = \sum_{k=0}^{\infty} f_k \in \hat{A}_k$ be given. We than inductively determine $g_k \in (S^kM)^*$ such that $\tilde{\varphi}\left(\sum_{k=0}^{\infty} g_k\right) = f$. Indeed the induction step to pass from g_k to g_{k+1} consists of solving the equaiton

$$\left(S^{k+1}T|K\right)^* g_{k+1} = f_{k+1} + r_{k+1}$$

where r_{k+1} is a function of $\varphi_1, \varphi_2, \ldots, \varphi_k, f_1, \ldots, f_k$ and $g_1, \ldots, g_k$. Thus we have shown that $\varphi|K : \hat{A}_M \longrightarrow \hat{A}_K$ is onto. Now let $f \in \hat{\mathfrak{m}}_L$. Then there exists $g \in \hat{\mathfrak{m}}_M$ such that $f - \tilde{\varphi}(g)$ vanishes on K. Hence there exists $h \in (\alpha_1, \ldots, \alpha_n)$ such that $f \equiv h \bmod I$. Thus there exist $c_1, c_2, \ldots, c_n \in \mathbf{k}$ such that $f \equiv c_1\alpha_1 + \cdots + c_n\alpha_n \bmod \hat{\mathfrak{m}}_L^2 + I$. But $\mathfrak{m}/\mathfrak{m}^2 \cong \hat{\mathfrak{m}}_L/(\hat{\mathfrak{m}}_L^2 + I)$ and the claim follows. Hence R_L is *Noetherian* by [Bo1], III, §10, Cor. 5. We let $\xi_n \in Y(R_L/\mathfrak{m}^n)$ be the element such that $e(\xi_n)$ is the projection $R_L \longrightarrow R_L/\mathfrak{m}^n$, $n \geq 1$. We then define $\xi \in Y(R_L)$ by $\xi = \varprojlim \xi_n$. Finally note that any algebra homomorphism $R_L \longrightarrow A$, $A \in \text{Obj } \mathcal{A}$, is continuous. □

Corollary. *Y_L is pro-representable.*

Proof. It remains to check that the kernel of $d\varphi(0) : C^1 \longrightarrow L^2$ is finite dimensional where $\varphi : C^1 \longrightarrow L^2$ is given by $\varphi(\eta) = d\eta + \frac{1}{2}[\eta, \eta]$. But clearly $d\varphi(0) = d$ whence

$$\operatorname{Ker} d\varphi(0) = Z^1(L) \cap C^1.$$

Clearly the projection $p : Z^1(L) \longrightarrow H^1(L)$ induces an isomorphism $\ker d\varphi(0) \longrightarrow H^1(L)$. □

We now address the problem of showing that the functor Iso $\mathcal{C}(L; \cdot)$ has a hull. We observe that the inclusion of functors $Y_L \longrightarrow$ Obj $\mathcal{C}(L; \cdot)$ induces a natural transformation of functors $\iota : Y_L \longrightarrow$ Iso $\mathcal{C}(L; \cdot)$. It suffices to show that ι is minimally smooth.

We begin by nothing that ι is induced by a homomorphism of differential graded Lie algebras. Indeed if we define $L_{\text{red}} \subset L$ by

$$L_{\text{red}} = C^1 \oplus \bigoplus_{i \geq 2} L^i$$

then $Y_L(A) = \text{Obj } \mathcal{C}(L_{\text{red}}; A) = \text{Iso } \mathcal{C}(L_{\text{red}}; A)$. If $i : L_{\text{red}} \longrightarrow L$ is the inclusion, then ι coincides with the natural transformation induced by i, $i : \text{Iso } \mathcal{C}(L_{\text{red}}; \cdot) \longrightarrow \text{Iso } \mathcal{C}(L; \cdot)$. We observe that i induces an isomorphism on H^1 and H^2 (but not on H^0).

Lemma 1.7. *Suppose $\phi : L \longrightarrow \overline{L}$ is a homomorphism of differential graded Lie algebras such that $H^1(\phi)$ is surjective and $H^2(\phi)$ is injective. Then the induced natural transformation $\phi : \text{Iso } \mathcal{C}(L; \cdot) \longrightarrow \text{Iso } (\overline{L}; \cdot)$ is smooth.*

Proof. We prove the lemma by Artinian induction. Let A, $\overline{A}$, $\mathfrak{m}$, $\mathfrak{J}$ be as above and $a \in \text{Obj } \mathcal{C}(L; \overline{A})$, $b \in \mathcal{C}(\overline{L}; \overline{A})$ with $\phi(a) = b$ and suppose there exists $\tilde{b} \in \text{Obj } \mathcal{C}(\overline{L}; A)$ lying over b. Let $o_2(a)$ be the obstruction to lifting a to Obj $\mathcal{C}(L; A)$, [GM1], §2. Then

$$H^2(\phi)\left(o_2(a)\right) = o_2\left(\phi(a)\right) = o_2(b) = 0.$$

Since $H^2(\phi)$ is injective, we conclude that there exists $\tilde{a} \in \text{Obj}\, \mathcal{C}(L; A)$ lying over a.

It remains to prove that we can choose the lift $\tilde{a}$ such that the images of $\phi(\tilde{a})$ and $\tilde{b}$ in Iso $\mathcal{C}(\overline{L}; A)$ agree. Now with $\tilde{a}$ as above we find that $\phi(\tilde{a})$ and $\tilde{b}$ are elements of Obj $\mathcal{C}(L;A)$ lying over b. Hence by [GM1], Proposition 2.6, there exists $c \in Z^1(\overline{L} \otimes \mathfrak{J})$ such that

$$\tilde{b} - \phi(\tilde{a}) = c.$$

Since $H^1(\phi)$ is surjective, we may choose $h \in Z^1(L \otimes \mathfrak{J})$ and $k \in \overline{L} \otimes \mathfrak{J}$ such that $c = \phi(h) + dk$ or

$$\tilde{b} - dk = \phi(\tilde{a} + h).$$

Now $\tilde{a} + h \in$ Obj $\mathcal{C}(L;A)$ again by [GM1], Proposition 2.6, and $\tilde{a} + h$ lies over a. Also $k \in \overline{L}^0 \otimes \mathfrak{J}$ and we have [GM1], Lemma 2.8,

$$\alpha(e^k)\tilde{b} = \tilde{b} - dk.$$

We obtain

$$\phi(\tilde{a} + h) = \alpha(e^k)\tilde{b}$$

and the lemma is proved. □

Corollary. $\iota : Y_L \longrightarrow$ Iso $\mathcal{C}(L;\cdot)$ *is smooth.*

We now want to prove that ι is *minimally* smooth.

Lemma 1.8. *Suppose* $A \in$ Obj $\mathcal{A}$ *satisfies* $\mathfrak{m}^2 = 0$. *Then*

$$\text{Iso } \mathcal{C}(L;A) = H^1(L \otimes \mathfrak{m}).$$

Proof. Since $\mathfrak{m}^2 = 0$, the graded Lie algebra $L \otimes \mathfrak{m}$ is abelian and

$$\text{Obj } \mathcal{C}(L;A) = Z^1(L \otimes \mathfrak{m}).$$

Also by [GM1], Lemma 2.7 (taking $\mathfrak{J} = \mathfrak{m}$), we obtain for $\lambda \in L^0 \otimes \mathfrak{m}$, $\eta \in L^1 \otimes \mathfrak{m}$

$$\alpha(e^\lambda) \cdot \eta = \eta - d\lambda.$$

The lemma follows. □

We have obtained the following theorem.

Theorem 1.3. *Suppose* $\phi : L \longrightarrow \overline{L}$ *is a homomorphism of differential graded Lie algebra such that* $H^1(\phi)$ *is an isomorphism and* $H^2(\phi)$ *is an injection. Then* $\phi :$ Iso $\mathcal{C}(L;\cdot) \longrightarrow$ Iso $\mathcal{C}(\overline{L};\cdot)$ *is minimally smooth.*

Corollary. $\iota : Y_L \longrightarrow$ Iso $\mathcal{C}(L;\cdot)$ *is minimally smooth.*

Corollary. *The functor* Iso $\mathcal{C}(L;\cdot)$ *has a hull* (R_L, ξ).

Proof. By Theorem 1.2 we have (R_L, ξ) such that $e_\xi : h_{R_L} \longrightarrow Y_L$ is an isomorphism of functors. Hence the composition $\iota \circ e_\xi : h_{R_L} \longrightarrow$ Iso $\mathcal{C}(L;\cdot)$ is minimally smooth. □

We can now make a definition which is of central importance to us.

Definition. Suppose $F : \mathcal{A} \longrightarrow$ **Sets** is a deformation functor and that there exists a differential graded Lie algebra L and a natural isomorphism of functors $\tau : Y_L \longrightarrow F$. Then we will say L is a *controlling differential graded Lie algebra* for F.

Remark. In the applications of the above general theory we will be given an analytic versal deformation $\pi : (E, e) \longrightarrow (X, x)$. Then we will say a differential graded Lie algebra L controls that deformation theory if there is an isomorphism $R_L \cong \hat{\mathcal{O}}_{X,x}$.

In order to prove the main theorem of this paper we will need to relate R_L for a certain differential graded Lie algebra to the completed local ring $\hat{\mathcal{O}}_{\mathcal{K}_M}$ of the "Kuranishi space" of M (for M a compact strongly pseudo-convex CR-manifold). We now give an abstract version of this. Theorem 1.4 was stated as Theorem 3.8 of [GM2] but the proof was not given there in detail. We now give the missing details. Let L be a differential graded Lie algebra.

We choose vector space splittings of the exact sequences, $0 \leq j < \infty$,

$$0 \longrightarrow Z^j(L) \longrightarrow L^j \xrightarrow{d} B^{j+1}(L) \longrightarrow 0$$

and

$$0 \longrightarrow B^j(L) \longrightarrow L^j \xrightarrow{d} H^j(L) \longrightarrow 0.$$

We let A^j, $j \geq 0$, denote the image of the splitting $B^{j+1} \longrightarrow L^j$ and H^j denote the image of the splitting $H^j(L) \longrightarrow Z^j(L)$. Using these splittings, we obtain maps $\delta : L^{j+1} \longrightarrow L^j$, [GM2], page 343, with $\delta^2 = 0$, such that the Hodge decomposition

$$d\delta + \delta d = I - H$$

holds. Here H is the projection onto the image of the cohomology under the above splitting.

Remark. In geometric situations, e.g. [K1] or [Mi1] we have $\delta = d^* G$ where G is the Green operator and H is the harmonic projection.

We define the Kuranishi map $F : L^1 \longrightarrow L^1$ by

$$F(\xi) = \xi + \tfrac{1}{2}\delta[\xi, \xi].$$

If $A \in \operatorname{Obj} \mathcal{A}$ has maximal ideal $\mathfrak{m}$ then we use F, δ, H and d to denote the induced maps on $L \otimes \mathfrak{m}$. We then have Lemma 3.11 of [GM2]:

Lemma 1.9. *$F : L^1 \otimes \mathfrak{m} \longrightarrow L^1 \otimes \mathfrak{m}$ is an isomorphism.*

We now define a new functor $\mathcal{K}_L : \mathcal{A} \longrightarrow$ **Sets** by

$$\mathcal{K}_L(A) = \left\{ \eta \in H^1 \otimes \mathfrak{m} : H\left([F^{-1}\eta, F^{-1}\eta]\right) = 0 \right\}.$$

Remark. In geometric situations $\mathcal{K}_L$ is pro-represented by the completed local ring of the Kuranishi space, page 350 of [GM2].

Our final theorem of this chapter is the following.

Theorem 1.4. *The Kuranishi map F induces a natural isomorphism of functors $F : Y_L \longrightarrow \mathcal{K}_L$.*

The proof of the theorem follows from the next three lemmas. Let $A \in \text{Obj } \mathcal{A}$.

Lemma 1.10. $F(Y_L(A)) \subset \mathcal{K}_L(A)$.

Proof. The proof is identical to that of [GM2], Lemma 2.4, with the change that $\xi \in L^1 \otimes \mathfrak{m}$ instead of $\widehat{L^1}$. □

Lemma 1.11. *Suppose $\xi \in L^1 \otimes \mathfrak{m}$ satisfies*

$$\delta d[\xi, \xi] = \delta[\delta d[\xi, \xi], \xi].$$

Then

$$\delta d[\xi, \xi] = 0.$$

Proof. Suppose $\delta d[\xi, \xi] \neq 0$. Then, since $\mathfrak{m}$ is nilpotent there exists k such that $\delta d[\xi, \xi] \in L^1 \otimes \mathfrak{m}^k$ and $\delta d[\xi, \xi] \notin L^1 \otimes \mathfrak{m}^{k+1}$. But $\xi \in L^1 \otimes \mathfrak{m}$ and consequently $\delta[\delta d[\xi, \xi], \xi] \in L^1 \otimes \mathfrak{m}^{k+1}$ whence $\delta d[\xi, \xi] \in L^1 \otimes \mathfrak{m}^{k+1}$. □

Lemma 1.12. $F^{-1}(\mathcal{K}_L(A)) \subset Y_L(A)$.

Proof. Suppose $\eta \in \mathcal{K}_L(A)$ and $\xi \in L^1 \otimes \mathfrak{m}$ satisfies $F(\xi) = \eta$ whence

$$\xi + \tfrac{1}{2}\delta[\xi, \xi] = \eta.$$

Since $\delta\eta = 0$, we obtain $\delta\xi = 0$. By definition $H\left[F^{-1}\eta, F^{-1}\eta\right] = H[\xi, \xi] = 0$. Since η is closed we have

$$d\xi + \tfrac{1}{2}\delta d[\xi, \xi] = 0.$$

Thus the lemma is proved if we can establish $d\delta[\xi, \xi] = [\xi, \xi]$. By the Hodge decomposition we have

$$d\delta[\xi, \xi] = -\delta d[\xi, \xi] + [\xi, \xi].$$

Thus it suffices to prove $\delta d[\xi, \xi] = 0$. But by the equation above

$$\delta d[\xi, \xi] = 2\delta[d\xi, \xi] = -\delta[d\delta[\xi, \xi], \xi].$$

Again using the Hodge decomposition and the formula $[[\xi, \xi], \xi] = 0$ (a consequence of the Jacobi identity) we find

$$\delta d[\xi, \xi] = \delta\left[\delta d[\xi, \xi], \xi\right].$$

□

Remark. The proof of Theorem 1.4 uses only the Hodge decomposition in L^2.

2. Vector-Valued Differential Forms on Complex Manifolds

In this section we will review the results of Frölicher and Nijenhuis connecting differential forms with values in the tangent bundle (or the complexified tangent bundle) with graded derivations of the de Rham algebra which we will use in what follows and prove analogous results for the space of graded derivations of the Dolbeault algebra.

Let M be a differentiable manifold of dimension n. Then the graded vector space $\mathcal{A}^{\bullet}(M, T(M) \otimes \mathbb{C}) = \bigoplus_{k=0}^{n} \mathcal{A}^{k}(M, T(M) \otimes \mathbb{C})$ of differential forms with values in the complexified tangent bundle becomes a graded Lie algebra when equipped with the Nijenhuis bracket defined as follows. Suppose $\deg \varphi = p$ and $\deg \psi = q$ and $X_1, X_2, \ldots, X_{p+q}$ are locally-defined complex vector fields. Then $[\varphi, \psi]$ is given by the formula of Nijenhuis

$$
\begin{aligned}
&[\varphi, \psi](X_1, X_2, \ldots, X_{p+q}) = \\
&\quad \tfrac{1}{p!q!} \sum_{\sigma} \varepsilon(\sigma) \left[\varphi\left(X_{\sigma(1)}, \ldots, X_{\sigma(p)}\right), \psi\left(X_{\sigma(q+1)}, \ldots, X_{\sigma(p+q)}\right)\right] + \\
&\quad \tfrac{(-1)^{pq+q+1}}{(p-1)!q!} \sum_{\sigma} \varepsilon(\sigma) \varphi\left(\left[X_{\sigma(1)}, \psi\left(X_{\sigma(2)}, \ldots, X_{\sigma(q+1)}\right)\right], X_{\sigma(q+2)}, \ldots, X_{\sigma(p+q)}\right) + \\
&\quad \tfrac{(-1)^{p}}{p!(q-1)!} \sum_{\sigma} \varepsilon(\sigma) \psi\left(\left[X_{\sigma(1)}, \varphi\left(X_{\sigma(2)}, \ldots, X_{\sigma(p+1)}\right)\right], X_{\sigma(p+2)}, \ldots, X_{\sigma(p+q)}\right) + \\
&\quad \tfrac{(-1)^{pq+q}}{2(p-1)!(q-1)!} \sum_{\sigma} \varepsilon(\sigma) \varphi\left(\psi\left(\left[X_{\sigma(1)}, X_{\sigma(2)}\right], \ldots, X_{\sigma(q+1)}\right), X_{\sigma(q+2)}, \ldots, X_{\sigma(p+q)}\right) + \\
&\quad \tfrac{(-1)^{p+1}}{2(p-1)!(q-1)!} \sum_{\sigma} \varepsilon(\sigma) \psi\left(\varphi\left(\left[X_{\sigma(1)}, X_{\sigma(2)}\right], \ldots, X_{\sigma(p+1)}\right), X_{\sigma(p+2)}, \ldots, X_{\sigma(p+q)}\right).
\end{aligned}
$$

Here σ runs over the full symmetric group S_{p+q}.

Remarks. The reader will verify that $[\varphi, \psi]$ is multilinear over the functions with respect to the variables $X_1, X_2, \ldots, X_{p+q}$. Given φ we will often use $\mathcal{L}_{\varphi}$ to denote the graded derivation of $\mathcal{A}^{\bullet}(M, T(M) \otimes \mathbb{C})$ given by

$$\mathcal{L}_{\varphi}(\psi) = [\varphi, \psi].$$

There is another important bilinear operation on the vector-valued forms, that of interior multiplication or contraction. Let φ and ψ be as above. Then the interior product or contraction of ψ by φ will be denoted $\iota_{\varphi}\psi$ (or $\varphi \lrcorner \psi$). It is of degree $p+q-1$ and is defined by

$$
\begin{aligned}
&\iota_{\varphi}\psi(X_1, X_2, \ldots, X_{p+q-1}) = \\
&\qquad \frac{1}{p!(q-1)!} \sum_{\sigma} \varepsilon(\sigma) \psi\left(\varphi\left(X_{\sigma(1)}, \ldots, X_{\sigma(p)}\right), X_{\sigma(p+1)}, \ldots, X_{\sigma(p+q-1)}\right).
\end{aligned}
$$

Remark. In [FN1] this operation is denoted $\psi \bar{\wedge} \varphi$.

We will also need to define $\mathcal{L}_{\varphi}$ and ι_{φ} as operators on scalar-valued forms. Let φ be a vector-valued p–form and ψ be a scalar-valued q–form. Then

$\mathcal{L}_\varphi\psi\,(X_1, X_2, \ldots, X_{p+q})$ is defined by the same formula as in the case that ψ was vector-valued except that the first term is replaced by

$$\frac{1}{p!q!}\sum_\sigma \varepsilon(\sigma)\varphi\left(X_{\sigma(1)}, \ldots, X_{\sigma(p)}\right) \cdot \psi\left(X_{\sigma(p+1)}, \ldots, X_{\sigma(p+q)}\right)$$

and the second and fourth terms are replaced by zero. The form $\iota_\varphi\psi$ is defined by the same formula as before. Then $\mathcal{L}_\varphi$ and ι_φ are graded derivations of the algebra of scalar forms of degree p and $p-1$ respectively.

Most of the familiar relations involving $\mathcal{L}_X$ and ι_X as operators on scalar-valued differential forms when X is a vector field have analogues when X is a vector-valued form and $\mathcal{L}_X$ and ι_X operator on scalar-valued forms as described above. We record some of these in the next lemma. Let λ, μ, η be vector-valued forms of degrees ℓ, m and n respectively. Then we have the following lemma.

Lemma 2.1. *The graded brackets of the above operators on scalar-valued forms satisfy*

(VF1) $[\mathcal{L}_\lambda, \mathcal{L}_\mu] = \mathcal{L}_{[\lambda,\mu]}$.
(VF2) $[\iota_\lambda, \iota_\mu] = \iota_{\iota_\lambda\mu} - (-1)^{(\ell-1)(m-1)}\iota_{\iota_\mu\lambda}$.
(VF3) $[\iota_\lambda, \mathcal{L}_\mu] = \mathcal{L}_{\iota_\lambda\mu} + (-1)^m\, \iota_{[\lambda,\mu]}$.
(VF4) $[\iota_\eta\lambda, \mu] + (-1)^{\ell(n-1)}[\lambda, \iota_\eta\mu] - \iota_n[\lambda,\mu] = (-1)^{m(\ell-1)}\iota_{[\eta,\mu]}\lambda + (-1)^{(\ell-1)}\iota_{[\eta,\lambda]}\mu$.

In case certain interior products are zero these formulas become

(VF3′) $[\iota_\lambda, \mathcal{L}_\mu] = (-1)^m\iota_{[\lambda,\mu]}$, *if* $\iota_\lambda\mu = 0$.
(VF4′) $\iota_\eta[\lambda,\mu] = (-1)^{m(\ell-1)+1}\iota_{[\eta,\mu]}\lambda + (-1)^\ell\iota_{[\eta,\lambda]}\mu$, *if* $\iota_\eta\lambda = \iota_\eta\mu = 0$.

Remarks. The formula (VF1) states that the map $\Phi : \mathcal{A}^\cdot(M, T(M)\otimes\mathbb{C}) \longrightarrow \operatorname{Der}\mathcal{A}^\cdot(M)$ given by $\Phi(\lambda) = \mathcal{L}_\lambda$ preserves brackets. In fact this is the point of the Nijenhuis formula for the bracket of two vector-valued forms. Recall that $\mathcal{A}^\cdot(M)$ denotes the (complex) de Rham algebra. It is amusing to note that the exterior differential $d \in \operatorname{Der}^1\mathcal{A}^\cdot(M)$ satisfies $d = \Phi(id)$, note $id = id_{T(M)\otimes\mathbb{C}} \in \mathcal{A}^1(M, T(M)\otimes\mathbb{C})$.

We record the special formulas for brackets and contractions of decomposables $\varphi\otimes X$ and $\psi\otimes Y$ with φ and ψ scalar forms of degrees p and q respectively and X and Y vector fields. The proofs are left to the reader.

Lemma 2.2. (i) $[\varphi\otimes X, \psi] = \varphi\wedge\mathcal{L}_X(\psi) + (-1)^p d\varphi\wedge\iota_X\psi$.

(ii) $(\varphi\otimes X)\lrcorner\,\psi = \varphi\wedge\iota_X(\psi)$. □

Lemma 2.3. (i) $[\varphi\otimes X, \psi\otimes Y] = \varphi\wedge\psi\otimes[X,Y] + [\varphi\otimes X, \psi]\otimes Y - (-1)^{pq}[\psi\otimes Y, \varphi]\otimes X$.

(ii) $(\varphi\otimes X)\lrcorner\,(\psi\otimes Y) = \varphi\wedge(\iota_X\psi)\otimes Y$. □

Finally we record the generalized Cartan formula, an immediate consequence of formula (VF3) and the formula $d = \mathcal{L}_{id}$.

Lemma 2.4. *Let φ be a vector-valued form of degree p. Then in* $\operatorname{Der}\mathcal{A}^{\cdot}(M)$ *we have*

$$[\iota_\varphi, d] = \mathcal{L}_\varphi. \qquad \square$$

In case M is a complex manifold we can decompose the vector-valued forms according to Hodge type and according to the decomposition

$$T(M)\otimes\mathbb{C} = T^{1,0}(M)\oplus T^{0,1}(M).$$

We wish to prove that the graded subspace $\mathcal{L}^{\cdot}(M)\subset\mathcal{A}^{\cdot}\left(M, T^{1,0}(M)\right)$ defined by $\mathcal{L}^p(M) = \mathcal{A}^{o,p}\left(M, T^{1,0}(M)\right)$ is closed under the Nijenhuis bracket. We also wish to show that the natural $\bar{\partial}$–operator on $\mathcal{L}^{\cdot}(M)$ is a graded derivation of $[\ ,\ ]$, thus we obtain a differential graded Lie algebra $\left(\mathcal{L}^{\cdot}(M), [\ ,\ ], \bar{\partial}\right)$. We accomplish both these goals by proving three general lemmas. These lemmas may be used to show that for a normal CR-manifold Kuranishi's complex $(\mathcal{K}^{\cdot}, \bar{\partial}_b)$, see Chapter 3, may be given a canonical differential graded Lie algebra structure, see [M2].

Lemma 2.5. *Let A be an involutive complex distribution on a manifold M. Then the subspace $\mathcal{A}$ of A–valued forms φ on M satisfying*

$$\iota_Z\varphi = 0, \quad \textit{all } Z\in A,$$

is closed under the Nijenhuis bracket of vector-valued forms.

Proof. We first observe that from the defining formula for the Nijenhuis bracket it is obvious that $[\varphi,\psi]$ takes values in A if φ and ψ do.

It remains to prove that $[\varphi,\psi]$ satisfies

$$\iota_Z[\varphi,\psi] = 0, \quad \text{all } Z\in A.$$

Clearly if $\varphi,\psi\in\mathcal{A}$ then $\iota_\varphi\psi = 0$ so we may apply (VF4′) above with $\lambda=\varphi$, $\mu=\psi$, $\eta = Z$ to obtain

$$\iota_Z[\varphi,\psi] = (-1)^{q(p-1)+1}\iota_{[Z,\psi]}\varphi + (-1)^p\iota_{[Z,\varphi]}\psi\ .$$

Since A is involutive the forms $[Z,\psi]$ and $[Z,\varphi]$ again take values in A and the lemma follows. $\square$

We now want to make $\mathcal{A}$ into a *differential* graded Lie algebra. Assume then that we have a splitting

$$T(M)\otimes\mathbb{C} = A\oplus B$$

where both A and B are involutive. Let P and Q be the projections on A and B respectively.

Lemma 2.6. $[Q,Q]=0.$

Proof. By the Nijenhuis formula we have for X, Y locally defined sections of $T(M)\otimes\mathbb{C}$:

$$[Q,Q](X,Y) = 2\,[Q(X),Q(Y)] - 2Q\left([X,Q(Y)]\right) + 2Q\left([Y,Q(X)]\right) + 2Q\left(Q[X,Y]\right).$$

Since $[Q,Q]$ is bilinear it suffices to examine the three cases

(i) $X,Y\in\Gamma(A)$.
(ii) $X\in\Gamma(A),\quad Y\in\Gamma(B)$.

(iii) $X, Y \in \Gamma(B)$.

The vanishing of the right-hand side for each of (i), (ii) and (iii) is easily verified. □

We now define a differential $D : \mathcal{A} \longrightarrow \mathcal{A}$ by

$$D\varphi = [Q, \varphi].$$

Lemma 2.7. *D carries $\mathcal{A}$ into itself.*

Proof. We first check that $D\varphi = [Q, \varphi]$ is A–valued. We have by (VF3)

$$(-1)^p \iota_{[\varphi,Q]} Q = [\iota_\varphi, \mathcal{L}_Q]Q - [\iota_\varphi Q, Q].$$

The right-hand side is clearly zero since $\mathcal{L}_Q Q = [Q, Q] = 0$. Next we prove

$$\iota_Z[Q, \varphi] = 0, \quad \text{all } Z \in A.$$

We have just seen that the vector-valued 1-forms $[Z, Q]$ and $[Q, \varphi]$ take values in A. We then apply (VF4′) with $\lambda = Q$, $\mu = \varphi$ and $\eta = Z$ to deduce

$$-\iota_Z[Q, \varphi] = \iota_{[Z,\varphi]} Q + \iota_{[Z,Q]} \varphi = 0.$$

□

We have obtained the following theorem.

Theorem 2.1. *Suppose we are given a splitting*

$$T(M) \otimes \mathbb{C} = A \oplus B$$

of $T(M) \otimes \mathbb{C}$ into the direct sum of involutive distributions. Let $\mathcal{A}$ be the subspace of vector-valued forms satisfying

(i) *φ takes values in A.*
(ii) *$\iota_Z \varphi = 0$, all $Z \in A$.*

Let Q be the projection on B with kernel A. Then $(\mathcal{A}, [\ ,\ ], \mathcal{L}_Q)$ is a differential graded Lie algebra. □

Example. Let M be a complex manifold of complex dimension n. Let $A = T^{1,0}(M)$, $B = T^{0,1}(M)$ so we obtain an involutive splitting of $T(M) \otimes \mathbb{C}$ and by the above theorem a differential graded Lie algebra with

$$\mathcal{L}(M) = \bigoplus_{q=0}^{n} \mathcal{A}^{0,q}\left(M, T^{1,0}(M)\right).$$

The reader will verify that $\mathcal{L}_Q = \bar{\partial}$ in this case and will note that if $\varphi \in \mathcal{L}^p(M)$, $\psi \in \mathcal{L}^q(M)$ then in the Nijenhuis formula in the beginning of this section the last two terms are zero.

Definition. The differential graded Lie algebra $\left(\mathcal{L}^\bullet(M), [\ ,\ ], \bar{\partial}\right)$ will be called the *Kodaira-Spencer algebra*. We will usually denote it $\mathcal{L}(M)$.

We let $\mathcal{A}^{0,\cdot}(M)$ denote the Dolbeault algebra (the differential graded algebra of exterior differential forms with holomorphic degree zero equipped with $\bar{\partial}$). The reader will verify that if $\omega \in \mathcal{L}(M)$ then $\mathcal{L}_\omega$ carries $\mathcal{A}^{0,\cdot}(M)$ into itself, thus we have a map $\Phi : \mathcal{L}(M) \longrightarrow \operatorname{Der} \mathcal{A}^{0,\cdot}(M)$ given by

$$\Phi(\omega) = \mathcal{L}_\omega.$$

The rest of this chapter will be devoted to proving the following theorem and the generalizations of it detailed below.

Theorem 2.2. *The map $\Phi : \mathcal{L}(M) \longrightarrow \operatorname{Der} \mathcal{A}^{0,\cdot}(M)$ is a quasi-isomorphism of differential graded Lie algebras.*

Remark. We will see below that Φ is not a surjection; however, there is a complementary acyclic complex to the image of Φ.

Lemma 2.8. *Φ is an injective homomorphism of differential graded Lie algebras.*

Proof. Clearly Φ is injective. Also from [FN1] we have

$$[\Phi(\omega_1), \Phi(\omega_2)] = [\mathcal{L}_{\omega_1}, \mathcal{L}_{\omega_2}] = \mathcal{L}_{[\omega_1, \omega_2]} = \Phi_{[\omega_1, \omega_2]}$$

whence Φ preserves brackets. Also if $Q : T(M) \otimes \mathbb{C} \longrightarrow T^{0,1}(M)$ is the projection, η is a scalar form and ω is a vector form we have by [FN2], (4.2),

$$\bar{\partial}\eta = \mathcal{L}_Q \eta$$

and by [FN2], (4.4), we have

$$\bar{\partial}\omega = [Q, \omega].$$

Hence

$$\Phi(\bar{\partial}\omega) = [\mathcal{L}_Q, \mathcal{L}_\omega] = \operatorname{ad} \bar{\partial}(\Phi(\omega)).$$

Thus Φ carries the differential of $\mathcal{L}^\cdot(M)$ into that of $\operatorname{Der} \mathcal{A}^{0,\cdot}(M)$. □

We warn the reader that Φ *is not onto*. There are two other families of graded derivations of $\mathcal{A}^{0,\cdot}(M)$ of degree p. The elements of the first family are interior multiplications ι_η by forms η of type $(0, p+1)$ with values in $T^{0,1}(M)$. We recall that ι_η is a graded derivation of degree p of the complexified de Rham algebra $\mathcal{A}^{0,\cdot}(M)$. Clearly ι_η carries the Dolbeault algebra into itself.

The second type of derivation of degree p of $\mathcal{A}^{0,\cdot}(M)$ is the sum of $\mathcal{L}_\eta$ and an interior multiplication depending on η with $\eta \in \mathcal{A}^{0,p}\left(M, T^{0,1}(M)\right)$. For any such η the derivation of $\mathcal{L}_\eta$ operates on $\mathcal{A}^\cdot(M)$ but does not necessarily carry the Dolbeault algebra into itself. We now derive a formula for a form $\mu = \mu(\eta)$ such that $\mathcal{A}_\eta = \mathcal{L}_\eta + \iota_\mu$ carries $\mathcal{A}^{0,\cdot}(M)$ into itself.

In order to do this we need to recall the graded derivation (of degree 1) $\mathcal{D}$ on the graded Lie algebra $\mathcal{A}^\cdot(M, T(M) \otimes \mathbb{C})$ of [FN2], page 551. We let $P : T(M) \otimes \mathbb{C} \longrightarrow T^{1,0}(M)$ and $Q : T(M) \otimes \mathbb{C} \longrightarrow T^{0,1}(M)$ be the projections whence P and Q are elements of $\mathcal{A}^1(M, T(M) \otimes \mathbb{C})$ such that $P + Q = id$. We also let $J = iP - iQ$ be the almost complex structure. Now it is easily checked that for any $T(M) \otimes \mathbb{C}$-valued form ω we have

$$[id, \omega] = 0$$

whence

$$[Q,\omega] = -[P,\omega] = \tfrac{i}{2}[J,\omega].$$

Following [FN2] we define a first order operator $\mathcal{D}$ on vector-valued forms by

$$\mathcal{D}\omega = [Q,\omega] = -[P,\omega].$$

It is proved in [FN2] that $\mathcal{D}^2 = 0$ and $\mathcal{D}$ carries $T^{0,1}(M)$–valued forms of type (r,s) into $T^{0,1}(M)$–valued forms of type $(r+1,s)$ and $T^{1,0}$–valued forms of type (r,s) into $T^{1,0}$–valued forms of type $(r,s+1)$.

Lemma 2.9. *Let $\gamma \in \mathcal{A}^{0,q}\left(M,T^{0,1}(M)\right)$. Define a graded derivation $\mathcal{A}_\gamma$ of degree q of $\mathcal{A}^{\cdot}\left(M,T(M)\otimes\mathbb{C}\right)$ by*

$$\mathcal{A}_\gamma = \mathcal{L}_\gamma - (-1)^{q+1}\iota_{\mathcal{D}\gamma}.$$

Then $\mathcal{A}_\gamma$ carries $\mathcal{A}^{0,\cdot}(M)$ into itself.

Proof. We first note that a linear endomorphism T of $\mathcal{A}^{\cdot}(M)$ carries $\mathcal{A}^{0,\cdot}(M)$ into itself if $[\iota_P,T]=0$ since $\mathcal{A}^{0,\cdot}(M) = \ker \iota_P$.

By the formula (VF3) of Lemma 2.1 we have

$$[\iota_P,\mathcal{L}_\gamma] = \mathcal{L}_{\iota_P\gamma} + (-1)^q\iota_{[P,\gamma]} = (-1)^{q+1}\iota_{\mathcal{D}\gamma}.$$

Also by (VF2)

$$[\iota_P,\iota_{\mathcal{D}\gamma}] = \iota_{\mathcal{D}\gamma}.$$

(Note that $\iota_P\omega = r\omega$, if and only if ω has holomorphic degree r.)

We obtain $[\iota_P,\mathcal{A}_\gamma]=0$ and the lemma follows. □

Lemma 2.10. *Let D be a derivation on $\mathcal{A}^{0,0}(M)$ with values in $\mathcal{A}^{0,q}(M)$. Then there exists a unique $\varphi \in \mathcal{A}^{0,q}\left(M,T(M)\otimes\mathbb{C}\right)$ such that*

$$D = \mathcal{L}_\varphi.$$

Proof. Clearly φ is unique if it exists. Let $Z_1, Z_2, \ldots, Z_q$ be smooth global sections of $T^{1,0}(M)$. We define a linear endomorphism $\delta : \mathcal{A}^{0,0}(M) \longrightarrow \mathcal{A}^{0,0}(M)$ by

$$\delta f = Df(\bar{Z}_1,\bar{Z}_2,\ldots,\bar{Z}_q).$$

Clearly δ is a derivation and consequently by a standard argument, [FN1], Proposition 3.3, there exists a complex vector field $V = V(\bar{Z}_1,\ldots,\bar{Z}_q)$ such that

$$\delta f = Vf.$$

We then define φ by the rule

$$\varphi(\bar{Z}_1,\ldots,\bar{Z}_q) = V(\bar{Z}_1,\ldots,\bar{Z}_q).$$

□

We now prove that there are no other graded derivations of $\mathcal{A}^{0,\cdot}(M)$ besides the ones described above.

Proposition 2.1. *Suppose D is a derivation of $\mathcal{A}^{0,\cdot}(M)$ of degree p. Then there exist unique $\alpha \in \mathcal{A}^{0,p}(M,T^{1,0})$, $\beta \in \mathcal{A}^{0,p}(M,T^{0,1})$ and $\gamma \in \mathcal{A}^{0,p+1}(M,T^{0,1}(M))$ such that*

$$D = \mathcal{L}_\alpha + \mathcal{A}_\beta + \iota_\gamma.$$

Proof. We first find α and β. The derivation D restricts to the ring of C^∞–functions to give a derivation D^0 on this ring with values in $\mathcal{A}^{0,p}(M)$. By the previous lemma there exists a vector-valued form φ so that $D^0 = \mathcal{L}_\varphi$. We decompose $\varphi = \alpha + \beta$ according to the decomposition

$$T(M) \otimes \mathbb{C} = T^{1,0}(M) \oplus T^{0,1}(M).$$

We extend D^0 to a derivation $\tilde{D}^0$ of $\mathcal{A}^{0,\cdot}(M)$ of degree p by the formula

$$\tilde{D}^0 = \mathcal{L}_\alpha + \mathcal{A}_\beta.$$

Thus the derivation $D' = D - \tilde{D}^0$ is a derivation of degree p that annihilates functions. Thus D' is linear over the functions and consequently is a cross-section of the bundle $\operatorname{Der}^p\left(\Lambda^\cdot T^{0,1}(M)^*\right)$. Any such cross-section is the (graded) derivation extension of the element of $\operatorname{Hom}\left(T^{0,1}(M)^*, \Lambda^{p+1}T^{0,1}(M)^*\right)$ obtained by restriction. Such a linear map coincides with contraction by a cross-section γ of $T^{0,1}(M) \otimes \Lambda^{p+1}T^{0,1}(M)^*$. Thus $D' = \iota_\gamma$ and the proposition follows. □

Before proving Theorem 2.2, we need one more lemma analyzing the effect of the differential $\operatorname{ad}\bar\partial$ in $\operatorname{Der}\mathcal{A}^{0,\cdot}(M)$ on the three types of elements of degree p considered above.

Lemma 2.11. *Let $\alpha \in \mathcal{A}^{0,p}\left(M,T^{1,0}(M)\right)$, $\beta \in \mathcal{A}^{0,p}\left(M,T^{0,1}(M)\right)$ and $\gamma \in \mathcal{A}^{0,p+1}\left(M,T^{0,1}(M)\right)$. Then*

(i) $[\bar\partial, \mathcal{L}_\alpha] = \mathcal{L}_{\mathcal{D}\alpha} = \mathcal{L}_{\bar\partial\alpha}$.
(ii) $[\bar\partial, \mathcal{A}_\beta] = 0$.
(iii) $[\bar\partial, \iota_\gamma] = (-1)^{p+1}\mathcal{A}_\gamma$.

Proof. (i) is (VF1) and the definition $[\bar\partial, \mathcal{L}_\gamma] = [\mathcal{L}_Q, \mathcal{L}_\gamma]$. Since (iii) implies (ii), it suffices to prove (iii).

(iii) We have

$$\begin{aligned}
[\bar\partial, \iota_\gamma] &= (-1)^{p+1}[\iota_\gamma, \bar\partial] = (-1)^{p+1}[\iota_\gamma, \mathcal{L}_Q] \\
&= (-1)^{p+1}\left(\mathcal{L}_{\iota_\gamma Q} - \iota_{[\gamma,Q]}\right) \\
&= (-1)^{p+1}\left(\mathcal{L}_\gamma - (-1)^{p+2}\iota_{[Q,\gamma]}\right) \\
&= (-1)^{p+1}\left(\mathcal{L}_\gamma - (-1)^{p+2}\iota_{\mathcal{D}\gamma}\right) \\
&= (-1)^{p+1}\mathcal{A}_\gamma.
\end{aligned}$$

□

We are now ready to prove Theorem 2.2. By Proposition 2.1 we have

$$\operatorname{Der}^p\left(\mathcal{A}^{0,\cdot}(M)\right) = \mathcal{L}^p(M) \oplus \mathcal{A}^{0,p}\left(M,T^{0,1}(M)\right) \oplus \mathcal{A}^{0,p+1}\left(M,T^{0,1}(M)\right).$$

From Lemma 2.2 we see that the graded subspace $S^\cdot(M)$ of $\operatorname{Der}\left(\mathcal{A}^{0,\cdot}(M)\right)$ defined so that $S^p(M)$ is the sum of the last two summands in the above formula is an acyclic complex complementary to $\mathcal{L}^{\bullet}(M)$. Theorem 2.2 follows immediately. We will see later that $S^{\bullet}(M)$ coincides with $\mathrm{Der}_{\mathcal{O}}(\mathcal{A}^{0,\bullet})$ (see next paragraph for notation).

In Chapter 7 we will need various refinements of Theorem 2.2. Let $\mathcal{O}$ be the sheaf of holomorphic functions on M and $\mathcal{A}^{0,\bullet}$ and $\mathcal{L}^{\bullet}$ be the sheaves associated to $\mathcal{A}^{0,\bullet}(M)$ and $\mathcal{L}^{\bullet}(M)$. We have an inclusion $\iota : \mathcal{O} \longrightarrow \mathcal{A}^{0,\bullet}$. We wish to study the induced map

$$\iota^* : \mathrm{Der}\,\mathcal{A}^{0,\bullet} \longrightarrow \mathrm{Der}(\mathcal{O}, \mathcal{A}^{0,\bullet}).$$

Since we have seen in Proposition 2.1 that all the elements of $\mathrm{Der}\,\mathcal{A}^{0,\bullet}(M)$ are local, we have a homomorphism $\sigma : \mathrm{Der}\,\mathcal{A}^{0,\bullet}(M) \longrightarrow \mathrm{Der}\,\mathcal{A}^{0,\bullet}$ inverting the natural map $\mathrm{Der}\,\mathcal{A}^{0,\bullet} \longrightarrow \mathrm{Der}\,\mathcal{A}^{0,\bullet}(M)$ whence a quasiisomorphism $\sigma \circ \Phi : \mathcal{L}(M) \longrightarrow \mathrm{Der}\,\mathcal{A}^{0,\bullet}$. We define $\Psi = \iota^* \circ \sigma \circ \Phi$ and obtain a commutative diagram

$$\begin{array}{ccc} \mathrm{Der}\,\mathcal{A}^{0,\bullet} & \xrightarrow{\ \iota^*\ } & \mathrm{Der}(\mathcal{O}, \mathcal{A}^{0,\bullet}) \\ & {\scriptstyle \sigma\circ\Phi}\nwarrow \quad \nearrow{\scriptstyle \Psi} & \\ & \mathcal{L}(M) & \end{array}$$

We note that Ψ is the map on global sections induced by a sheaf map

$$\lambda : \mathcal{L}^{\bullet} \longrightarrow \mathcal{D}er(\mathcal{O}, \mathcal{A}^{0,\bullet}).$$

Here $\mathcal{D}er(\mathcal{O}, \mathcal{A}^{0,\bullet})$ denotes the *sheaf* of graded derivations from $\mathcal{O}$ to $\mathcal{A}^{0,\bullet}$. □

Lemma 2.12. *Ψ is an isomorphism.*

Proof. It suffices to prove that λ is an isomorphism. To see this let U be a coordinate patch on M and $(z_1, z_2, \ldots, z_n)$ the corresponding holomorphic local coordinates. The domain and range of $\lambda(U)$ are both free modules over $C^{\infty}(U)$ on the basis (for the terms of degree p)

$$\left\{ d\bar{z}_{i_1} \wedge \cdots \wedge d\bar{z}_{i_p} \otimes \frac{\partial}{\partial z_i} : 1 \le i \le n,\ 1 \le i_1 < \cdots < i_p \le n \right\}$$

and $\lambda(U)$ is the identity relative to this basis. □

Theorem 2.2 (bis).

$$\iota^* : \mathrm{Der}\,\mathcal{A}^{0,\bullet} \longrightarrow \mathrm{Der}(\mathcal{O}, \mathcal{A}^{0,\bullet})$$

is a quasi-isomorphism.

Proof. In the diagram above $\sigma \circ \Phi$ is a quasi-isomorphism and Ψ is an isomorphism. □

Corollary. *The differential graded Lie algebra* $\mathrm{Der}_{\mathcal{O}}(\mathcal{A}^{0,\bullet})$ *is acyclic.*

Proof. Under the isomorphism from $\mathrm{Der}(\mathcal{A}^{0,\bullet})$ to $\mathrm{Der}\left(\mathcal{A}^{0,\bullet}(M)\right)$, the subalgebra $\mathrm{Der}_{\mathcal{O}}(\mathcal{A}^{0,\bullet})$ goes to the acyclic complement $S^{\bullet}(M)$ to $\mathcal{L}^{\bullet}(M) \subset \mathrm{Der}\left(\mathcal{A}^{0,\bullet}(M)\right)$ constructed above. □

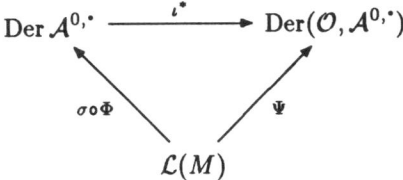

In Chapter 7 we will need a two-fold generalization of our computation of $\mathrm{Der}\left(\mathcal{A}^{0,\bullet}(M)\right)$. First we will need to allow coefficients in the graded symmetric algebra of a negatively-graded finite complex of holomorphic vector bundles on M. Second we will need to consider the relative case.

We consider the category $\mathcal{C}$ whose objects are triples $\widetilde{M} = (M, E^\bullet, \nabla)$ where M is a complex manifold, $E^\bullet = \left(\bigoplus_{\mu=-m}^{-1} E^\mu, s\right)$ is a cochain complex of holomorphic vector bundles over M and ∇ is a connection on $E^\bullet$ of type (1,0) which preserves the grading. We do not assume that the differential) s of $E^\bullet$ is parallel. We note that ∇ induces a connection, also denoted ∇, on the graded symmetric algebra. We require $\nabla(1) = 0$ where 1 is the constant function with value 1 on M considered as a section of $S^0(E^\bullet)$. A morphism $\tilde{p}_{\alpha\beta}$ in $\mathcal{C}$ from a triple $\widetilde{M}_\beta = (M_\beta, E^\bullet_\beta, \nabla_\beta)$ to a triple $\widetilde{M}_\alpha = (M_\alpha, E^\bullet_\alpha, \nabla_\alpha)$ is a pair, $(p_{\alpha\beta}, r_{\beta\alpha})$ were $p_{\alpha\beta} : M_\beta \longrightarrow M_\alpha$ is a holomorphic map and $r_{\beta\alpha} : p^*_{\alpha\beta}E^\bullet_\alpha \longrightarrow E^\bullet_\beta$ is a holomorphic map, linear on fibers, preserving grading and carrying the connection $p^*_{\alpha\beta}\nabla_\alpha$ to $\nabla_\beta \circ r_{\beta\alpha}$ and the differential $p^*_{\alpha\beta}s_\alpha$ to $s_\beta \circ r_{\beta\alpha}$.

Now let $\widetilde{M} = (M, E; \nabla)$ be as above. We define graded vector bundles $U^\bullet$ and $V^\bullet$ over M by

$$U^\bullet = T(M) \otimes \mathbb{C} \oplus T^{0,1}(M)[1] \oplus (E^\bullet)^*$$

and

$$V^\bullet = T^{1,0}(M) \oplus (E^\bullet)^*.$$

We put $\widetilde{E}^\bullet = T^{0,1}(M)^*[-1] \oplus E^\bullet$. Then

$$S(\widetilde{E}^\bullet) = \Lambda^\bullet T^{0,1}(M)^* \otimes S(E^\bullet).$$

If $p^*_{\beta\alpha} : \widetilde{M}_\beta \longrightarrow \widetilde{M}_\alpha$ is a morphism as above we define $\tilde{r}_{\beta\alpha} : p^*_{\alpha\beta}\widetilde{E}^\bullet_\alpha \longrightarrow \widetilde{E}^\bullet_\beta$ by $\tilde{r}_{\beta\alpha} = (dp_{\alpha\beta})^t \oplus r_{\beta\alpha}$.

Again let $\widetilde{M}$ be an object in $\mathcal{C}$. We define the differential bigraded algebra $\widetilde{R}^{\bullet,\bullet}(M)$ by

$$\widetilde{R}^{\bullet,\bullet}(M) = \left(\mathcal{A}^{0,\bullet}(M, S(E^\bullet)), \bar{\partial} \otimes 1,\ 1 \otimes s\right).$$

Here we define $\mathcal{A}^{0,\bullet}\left(M, S(E^\bullet)\right)$ by

$$\mathcal{A}^{0,\bullet}\left(M, S(E^\bullet)\right) = \Gamma\left(M, S(\widetilde{E}^\bullet)\right).$$

We will use $\deg(\mu)$ to denote the total degree of $\mu \in \widetilde{R}^{\bullet,\bullet}(M)$. We observe that a morphism $\tilde{p}_{\alpha\beta} : \widetilde{M}_\beta \longrightarrow \widetilde{M}_\alpha$ induces a bundle map $\tilde{p}^*_{\alpha\beta} : S(\widetilde{E}^\bullet_\alpha) \longrightarrow S(\widetilde{E}^\bullet_\beta)$ defined to be the composition $S(\tilde{r}_{\beta\alpha}) \circ p^*_{\alpha\beta}$. We denote the corresponding map on global sections by

$$\pi_{\beta\alpha} : \widetilde{R}^{\bullet,\bullet}(M_\alpha) \longrightarrow \widetilde{R}^{\bullet,\bullet}(M_\beta).$$

Our main interest in this section is to compute the space of graded derivations $\mathrm{Der}\left(\widetilde{R}^{\bullet,\bullet}(M_\alpha), \widetilde{R}^{\bullet,\bullet}(M_\beta)\right)$ in terms of differential forms on M_β with values in $p^*_{\alpha\beta}(U^\bullet_\alpha) \otimes S(E^\bullet_\beta)$.

Accordingly, we define extensions of the derivations above, again to be denoted, $\mathcal{L}_\mu$, $\mathcal{A}_\nu$, ι_φ, γ_ψ in $\mathrm{Der}^{p-r}\left(\widetilde{R}^{\cdot,\cdot}(M_\alpha), \widetilde{R}^{\cdot,\cdot}(M_\beta)\right)$ for

$$\mu \in \mathcal{A}^{0,p}\left(M_\beta, p_{\alpha\beta}^* T^{1,0}(M_\alpha) \otimes S(E_\beta^\cdot)^{-r}\right),$$
$$\in \mathcal{A}^{0,p}\left(M_\beta, p_{\alpha\beta}^* T^{0,1}(M_\alpha) \otimes S(E_\beta^\cdot)^{-r}\right),$$
$$\varphi \in \mathcal{A}^{0,p+1}\left(M_\beta, p_{\alpha\beta}^* T^{0,1}(M_\alpha) \otimes S(E_\beta^\cdot)^{-r}\right) \text{ and}$$
$$\psi \in \mathcal{A}^{0,p-q}\left(M_\beta, p_{\alpha\beta}^* (E_\alpha^*)^q \otimes S(E_\beta^\cdot)^{-r}\right).$$

The definitions of ι_φ and γ_ψ are the natural pointwise ones using the obvious contractions. We define

$$\mathcal{A}_\nu = \iota_\nu \circ \bar{\partial}_\alpha + (-1)^n \bar{\partial}_\beta \circ \iota_\nu, \quad \text{where } \deg(\nu) = n.$$

In order to define $\mathcal{L}_\mu$ we need the connections ∇_β and ∇_α. Let d_β (resp. d_α) be the exterior (covariant) differential on bundle-valued forms associated to ∇_β (resp. ∇_α). Let ∂_β (resp. ∂_α) be the $(1,0)$–part of d_β (resp. d_α). We then define

$$\mathcal{L}_\mu = \iota_\mu \circ d_\alpha + (-1)^m d_\beta \circ \iota_\mu, \quad \text{where } \deg(\mu) = m.$$

We note that since ι_μ annihilates $\widetilde{R}^{\cdot,\cdot}(M_\alpha)$ we have $\mathcal{L}_\mu = \iota_\mu \circ d_\alpha$.

If $D \in \mathrm{Der}\left(\widetilde{R}^{\cdot,\cdot}(M_\alpha), \widetilde{R}^{\cdot,\cdot}(M_\beta)\right)$ is of the form $D = \mathcal{L}_\mu$ we will say D is a derivation of type $\mathcal{L}$. We will use the analogous terminology (i.e. types $\mathcal{A}$, ι and γ) for derivations of the other three kinds above. That the above formulas for $\mathcal{A}_\nu$ and $\mathcal{L}_\mu$ define derivations may be deduced from the following general principle. If $f : R \longrightarrow S$ is an algebra homomorphism then $\mathrm{Der}(R,S)$ is a module over the fibre product of $\mathrm{Der}(R)$ and $\mathrm{Der}(S)$ over $\mathrm{Der}(R,S)$. Now note that (d_β, d_α) and $(\bar{\partial}_\beta, \bar{\partial}_\alpha)$ are elements of such a fibre product with $R = \mathrm{Der}\left(\widetilde{R}^{\cdot,\cdot}(M_\alpha)\right)$ and $S = \mathrm{Der}\left(\widetilde{R}^{\cdot,\cdot}(M_\beta)\right)$.

We observe that there is a natural map

$$\varphi : \Gamma\left(M_\beta, p_{\alpha\beta}^* T(M_\alpha)\right) \longrightarrow \mathrm{Der}\left(C^\infty(M_\alpha), C^\infty(M_\beta)\right)$$

defined as follows. Let $x \in M_\beta$ and $y = p_{\alpha\beta}(x)$. A section X of $p_{\alpha\beta}^* T(M_\alpha)$ assigns an element $X(x)$ of $T_y(M_\alpha)$ to x. Let $f \in C^\infty(M_\alpha)$. Then we define

$$\varphi(X) f(x) = X(x) f(y).$$

The next lemma is Proposition 3.2.7 of [Sa]. (One does not need π to be a bundle map in Saunder's proof.)

Lemma 2.13. *The natural map φ is an isomorphism.*

The proof of the next proposition is analogous to that of Proposition 2.1. Namely, given $D \in \mathrm{Der}\left(\widetilde{R}^{\cdot,\cdot}(M_\alpha), \widetilde{R}^{\cdot,\cdot}(M_\beta)\right)$ there exist μ and ν as above such that $D - \mathcal{L}_\mu - \mathcal{A}_\nu$ is linear over $C^\infty(M_\alpha)$ hence is a sum of derivations of type ι and γ.

Proposition 2.2. *Let* $D \in \mathrm{Der}^{p-r}\left(\widetilde{R}^{\cdot,\cdot}(M_\alpha), \widetilde{R}^{\cdot,\cdot}(M_\beta)\right)$. *Then there exist unique* μ, ν, φ, ψ *as above such that*

$$D = \mathcal{L}_\mu + \mathcal{A}_\nu + \iota_\varphi + \gamma_\psi. \qquad \square$$

Since every D as above is local we obtain the sheaf-theoretic version of the previous proposition. Let $\widetilde{R}_\alpha^{\cdot,\cdot}$ (resp. $\widetilde{R}_\beta^{\cdot,\cdot}$) denote the sheaf corresponding to $\widetilde{R}^{\cdot,\cdot}(M_\alpha)$ (resp. $\widetilde{R}^{\cdot,\cdot}(M_\beta)$).

Corollary. *Let* $D \in \mathrm{Der}^{p-r}(\widetilde{R}_\alpha^{\cdot,\cdot}, \widetilde{R}_\beta^{\cdot,\cdot})$. *Then there exist unique* μ, ν, φ, ψ *as above such that*

$$D = \mathcal{L}_\mu + \mathcal{A}_\nu + \iota_\varphi + \gamma_\psi. \qquad \square$$

The next theorem is an immediate consequence of Proposition 2.2.

Theorem 2.3. (i) *There is a natural isomorphism*

$$\Phi_{\beta\alpha} : \Gamma\left(M_\beta, \mathrm{Hom}\left(p_{\alpha\beta}^*(U_\alpha^\cdot)^*, S(\widetilde{E}_\beta^\cdot)\right)\right) \longrightarrow \mathrm{Der}\left(\widetilde{R}^{\cdot,\cdot}(M_\alpha), \widetilde{R}^{\cdot,\cdot}(M_\beta)\right).$$

(ii) *Under the isomorphisms* $\Phi_{\alpha\alpha}$ *and* $\Phi_{\beta\alpha}$ *the map*

$$(\pi_{\beta\alpha})_* : \mathrm{Der}\left(\widetilde{R}^{\cdot,\cdot}(M_\alpha)\right) \longrightarrow \mathrm{Der}\left(\widetilde{R}^{\cdot,\cdot}(M_\alpha), \widetilde{R}^{\cdot,\cdot}(M_\beta)\right)$$

corresponds to the map on global sections induced by the bundle map $\tilde{p}_{\alpha\beta}^* : S(\widetilde{E}_\alpha^\cdot) \longrightarrow S(\widetilde{E}_\beta^\cdot)$.

(iii) *Under the isomorphisms* $\Phi_{\beta\beta}$ *and* $\Phi_{\beta\alpha}$ *the map*

$$(\pi_{\beta\alpha})^* : \mathrm{Der}\left(\widetilde{R}^{\cdot,\cdot}(M_\beta)\right) \longrightarrow \mathrm{Der}\left(\widetilde{R}^{\cdot,\cdot}(M_\alpha), \widetilde{R}^{\cdot,\cdot}(M_\beta)\right)$$

corresponds to the map induced on global sections by the bundle map $p_{\alpha\beta}^*(U_\alpha^\cdot)^* \longrightarrow (U_\beta^\cdot)^*$. $\square$

We will need a result analogous to Theorem 2.3 for certain quotients of $\mathrm{Der}(\widetilde{R}_\alpha^{\cdot,\cdot}, \widetilde{R}_\beta^{\cdot,\cdot})$ obtained by restricting to a subalgebra of $\widetilde{R}_\alpha^{\cdot,\cdot}$. Let $\mathcal{E}_\alpha^\cdot$ denote the graded sheaf of holomorphic sections of $E_\alpha^\cdot$ and let $R_\alpha^\cdot$ be the differential graded $\mathcal{O}_{M_\alpha}$-algebra given by

$$R_\alpha^\cdot = S_{\mathcal{O}_{M_\alpha}}(\mathcal{E}_\alpha^\cdot).$$

We have an inclusion $R_\alpha^\cdot \subset \widetilde{R}_\alpha^{\cdot,\cdot}$. The reader will check that restricting a derivation D of $\widetilde{R}_\alpha^{\cdot,\cdot}$ to $R_\alpha^\cdot$ corresponds under the isomorphism of Proposition 2.2 to projecting onto the sum of the pieces of D of types $\mathcal{L}$ and γ. Thus we obtain the following theorem (here we let $\pi'_{\beta\alpha} : R_\alpha^\cdot \longrightarrow R_\beta^\cdot$ be the natural map over $p_{\alpha\beta}$).

Theorem 2.3 (bis). (i) *There is a natural isomorphism*

$$\Phi'_{\beta\alpha} : \Gamma\left(M_\beta, \mathrm{Hom}\left(p_{\alpha\beta}^*(V_\alpha^\cdot)^*, S(\widetilde{E}_\beta^\cdot)\right)\right) \longrightarrow \mathrm{Der}(R_\alpha^\cdot, \widetilde{R}_\beta^{\cdot,\cdot}).$$

(ii) *Under the isomorphisms* $\Phi'_{\alpha\alpha}$ *and* $\Phi'_{\beta\alpha}$ *the map*

$$(\pi_{\beta\alpha})_* : \mathrm{Der}(R^\cdot_\alpha, \widetilde{R}^{\cdot,\cdot}_\alpha) \longrightarrow \mathrm{Der}(R^\cdot_\alpha, \widetilde{R}^{\cdot,\cdot}_\beta)$$

corresponds to the map on global sections induced by the bundle map $\widetilde{p}^*_{\alpha\beta} : S(\widetilde{E}^\cdot_\alpha) \longrightarrow S(\widetilde{E}^\cdot_\beta)$.

(iii) *Under the isomorphisms* $\Phi'_{\beta\beta}$ *and* $\Phi'_{\alpha\beta}$ *the map*

$$(\pi'_{\beta\alpha})^* : \mathrm{Der}(R^\cdot_\beta, \widetilde{R}^{\cdot,\cdot}_\beta) \longrightarrow \mathrm{Der}(R^\cdot_\alpha, \widetilde{R}^{\cdot,\cdot}_\beta)$$

corresponds to the map induced on global sections by the bundle map $p^*_{\alpha\beta}(V^\cdot_\alpha)^* \longrightarrow (V^\cdot_\beta)^*$. □

Next we will need a generalization of Theorem 2.3 when we pass to the quotient of $\widetilde{R}^{\cdot,\cdot}_\alpha(M_\beta)$ determined by a subobject of $\widetilde{M}_\beta$. Let $\widetilde{M} = (M, E^\cdot, \nabla)$ be an object in $\mathcal{C}$. Then an object $\widetilde{N} = (N, \bar{E}^\cdot, \bar{\nabla})$ is a subobject of $\widetilde{M}$ if N is a complex submanifold of M, $\bar{E}^\cdot$ is a graded quotient of $E^\cdot|N$ and $\bar{\nabla}$ is induced by ∇. Now suppose we are given a morphism $\tilde{p}_{\alpha\beta} : \widetilde{M}_\beta \longrightarrow \widetilde{M}_\alpha$, subobjects $\widetilde{N}_\alpha \subset \widetilde{M}_\alpha$, $\widetilde{N}_\beta \subset \widetilde{M}_\beta$ and a morphism $\tilde{q}_{\alpha\beta} : \widetilde{N}_\beta \longrightarrow \widetilde{N}_\alpha$ compatible with $\tilde{p}_{\alpha\beta}$. We define $\bar{\widetilde{E}}^\cdot_\alpha = T^{0,1}(N_\alpha)^*[-1] \oplus \bar{E}^\cdot_\alpha$ and define $\bar{\widetilde{E}}^\cdot_\beta$ analogously. We observe that we have natural maps $\widetilde{R}^{\cdot,\cdot}(M_\alpha) \longrightarrow \widetilde{R}^{\cdot,\cdot}(N_\alpha)$ and $\widetilde{R}^{\cdot,\cdot}(M_b) \longrightarrow \widetilde{R}^{\cdot,\cdot}(N_\beta)$.

Theorem 2.3 (tertio). (i) *There is a natural isomorphism*

$$\bar{\Phi}_{\beta\alpha} : \Gamma\left(M_\beta, \mathrm{Hom}\left(p^*_{\alpha\beta}(U^\cdot_\alpha)^*|N_\beta, S(\bar{\widetilde{E}}^\cdot_\beta)\right)\right) \longrightarrow \mathrm{Der}\left(\widetilde{R}^{\cdot,\cdot}(M_\alpha), \widetilde{R}^{\cdot,\cdot}(N_\beta)\right).$$

(ii) *Under the isomorphisms* $\bar{\Phi}_{\alpha\alpha}$ *and* $\bar{\Phi}_{\beta\alpha}$ *map*

$$(\bar{\pi}_{\beta\alpha})_* : \mathrm{Der}\left(\widetilde{R}^{\cdot,\cdot}(M_\alpha), \widetilde{R}^{\cdot,\cdot}(N_\alpha)\right) \longrightarrow \mathrm{Der}\left(\widetilde{R}^{\cdot,\cdot}(M_\alpha), \widetilde{R}^{\cdot,\cdot}(N_\beta)\right)$$

corresponds to the map on global sections induced by the bundle map $\tilde{q}^*_{\alpha\beta} : S(\bar{\widetilde{E}}^\cdot_\alpha) \longrightarrow S(\bar{\widetilde{E}}^\cdot_\beta)$.

(iii) *Under the isomorphisms* $\bar{\Phi}_{\beta\beta}$ *and* $\bar{\Phi}_{\beta\alpha}$ *the map*

$$\pi^*_{\beta\alpha} : \mathrm{Der}\left(\widetilde{R}^{\cdot,\cdot}(M_\beta), \widetilde{R}^{\cdot,\cdot}(N_\beta)\right) \longrightarrow \mathrm{Der}\left(\widetilde{R}^{\cdot,\cdot}(M_\alpha), \widetilde{R}^{\cdot,\cdot}(N_\beta)\right)$$

corresponds to the map induced on global sections by the bundle map $p^*_{\alpha\beta}(U^\cdot_\alpha)^* \longrightarrow (U^\cdot_\beta)$.

Proof. The statement (i) follows from Theorem 2.3(i) with M_β replaced by N_β and $p_{\alpha\beta}$ by $p_{\alpha\beta} \circ i_\beta$ where $i_\beta : N_\beta \longrightarrow M_\beta$ is the inclusion.

In order to prove (ii) and (iii) consider a pair of morphisms $\tilde{p}_{\beta\gamma} : \widetilde{M}_\gamma \longrightarrow \widetilde{M}_\beta$ and $\tilde{p}_{\alpha\beta} : \widetilde{M}_\beta \longrightarrow \widetilde{M}_\alpha$. We then have induced maps as in Theorem 2.3(i)

$$(\pi_{\alpha,\gamma\beta})_* : \mathrm{Der}\left(\widetilde{R}^{\cdot,\cdot}(M_\alpha), \widetilde{R}^{\cdot,\cdot}(M_\beta)\right) \longrightarrow \mathrm{Der}\left(\widetilde{R}^{\cdot,\cdot}(M_\alpha), \widetilde{R}^{\cdot,\cdot}(M_\gamma)\right)$$

and

$$(\pi_{\beta\alpha,\gamma})^* : \mathrm{Der}\left(\widetilde{R}^{\bullet,\bullet}(M_\beta), \widetilde{R}^{\bullet,\bullet}(M_\gamma)\right) \longrightarrow \mathrm{Der}\left(\widetilde{R}^{\bullet,\bullet}(M_\alpha), \widetilde{R}^{\bullet,\bullet}(M_\gamma)\right).$$

It is easy to check that these maps are induced by the natural bundle maps. Then (ii) is a special case of the first formula above with M_β replaced by N_α and M_γ by N_β. Finally (iii) is the special case of the above with M_γ replaced by N_β. □

Finally we will need the corresponding generalization of Theorem 2.3(bis). We let $\widetilde{R}^{\bullet,\bullet}_\alpha$ (resp. $\widetilde{R}^{\bullet,\bullet}_\beta$) denote the sheaves of C^∞–sections of $S(\widetilde{\widetilde{E}}^\bullet_\alpha)$ (resp. $S(\widetilde{\widetilde{E}}^\bullet_\beta)$). We leave the proof of the next theorem to the reader.

Theorem 2.3 (quarto). (i) *There is a natural isomorphism*

$$\bar{\Phi}'_{\beta\alpha} : \Gamma\left(M_\beta, \mathrm{Hom}\left(p^*_{\alpha\beta}(V^\bullet_\alpha)^* \mid N_\beta, S(\widetilde{\widetilde{E}}^\bullet_\beta)\right)\right) \longrightarrow \mathrm{Der}(R_\alpha, \widetilde{R}^{\bullet,\bullet}_\beta).$$

(ii) *Under the isomorphisms* $\bar{\Phi}'_{\alpha\alpha}$ *and* $\bar{\Phi}_{\beta\alpha}$ *the map*

$$(\bar{\pi}_{\beta\alpha})_* : \mathrm{Der}(R^\bullet_\alpha, \widetilde{R}^{\bullet,\bullet}_\alpha) \longrightarrow \mathrm{Der}(R^\bullet_\alpha, \widetilde{R}^{\bullet,\bullet}_\beta)$$

corresponds to the map on global sections induced by the bundle map $\widetilde{q}^*_{\alpha\beta} : S(\widetilde{\widetilde{E}}^\bullet_\alpha) \longrightarrow S(\widetilde{\widetilde{E}}^\bullet_\beta)$.

(iii) *Under the isomorphisms* $\bar{\Phi}_{\beta\beta}$ *and* $\bar{\Phi}'_{\beta\alpha}$ *the map*

$$\pi^*_{\beta\alpha} : \mathrm{Der}(R^\bullet_\beta, \widetilde{R}^{\bullet,\bullet}_\beta) \longrightarrow \mathrm{Der}(R^\bullet_\alpha, \widetilde{R}^{\bullet,\bullet}_\beta)$$

corresponds to the map on global sections induced by the bundle map $p^*_{\alpha\beta}(V^\bullet_\alpha)^* \longrightarrow (V^\bullet_\beta)^*$.

3. Kuranishi's CR Deformation Theory

Let V be an analytic subset of $\mathbb{C}^N$ of dimension n which has a unique singular point which we assume is located at the origin. We let $U = V - \{0\}$ be the regular part of V. Let S_ε be a small sphere around the origin and put $M = S_\varepsilon \cap U$. Then M inherits a strongly pseudo-convex CR-structure from U. The horizontal distribution H on M is defined as follows. Let $x \in M$. Then

$$H_x = T_x(M) \cap J_x T_x(M).$$

Here J denotes the complex structure on U. Then H_x is J_x–invariant by definition and the pair (H, J) is a strongly pseudo-convex CR-structure on M. Let $T^{1,0}(M)$ be the bundle of $+i$–eigenspaces for J acting on the complexification of $H \otimes \mathbb{C}$. We let $T^{0,1}(M) = \overline{T^{1,0}(M)}$, the complex conjugate of $T^{1,0}(M)$. Then

$$T^{0,1}(M) = T^{0,1}(U) \mid M \cap (T(M) \otimes \mathbb{C}).$$

Thus $T^{0,1}(M)$ is an integrable sub-bundle of $T(M) \otimes \mathbb{C}$. A pair (H, J) as above on an odd-dimensional manifold M such that the $\pm i$–eigenspaces of J on $H \otimes \mathbb{C}$ are integrable constitutes an (abstract) CR-structure on M, see [Ta], Chapter I.

Let ρ be a defining function for M so $\rho : V \longrightarrow \mathbb{R}$ with $\rho^{-1}(0) = M$, $d\rho \mid T(M)$ nowhere zero and ρ negative on the inside of M in V. We let $M_t = \rho^{-1}(t)$, for $t \in \mathbb{R}$. Then M is strongly pseudo-convex means that the Hermitian form $L(\rho)$ on $T^{1,0}(M)$ given by

$$L(\rho)(Z, W) = \partial\bar{\partial}\rho(Z, \overline{W})$$

is positive definite at each point of M. Here Z and W are smooth sections of $T^{1,0}(M)$. In fact we will take $\rho = r^2 - \varepsilon$ whence $L(\rho)$ is positive definite on $T^{1,0}(U)$. We note that $\ker \partial\rho|\,(T^{1,0}(U)|M_t) = T^{1,0}(M_t)$ and $\ker \bar{\partial}\rho|\,(T^{0,1}(U)|M_t) = T^{0,1}(M_t)$.

We will need some auxiliary vector fields and 1-forms associated to ρ. We define $\theta = Jd\rho|T(M)$ where $Jd\rho(X) = -d\rho(JX)$ for $X \in T(M)$. Then $\ker\theta = H$. We now construct a smooth vector field T on M everywhere transverse to H.

Lemma 3.1. *There exists a unique vector field ξ of type $(1,0)$ on U such that*

(i) $\partial\rho(\xi) = 1$
(ii) $\iota_\xi \partial\bar{\partial}\rho = c\bar{\partial}\rho$, *for some smooth complex-valued function c on U.*

Proof. The lemma follows from the observation that the annihilator $T^{0,1}(M_t)^\perp$ of $T^{0,1}(M_t)$ in $T^{1,0}(U)|M_t$ for the bilinear form $\partial\bar{\partial}\rho$ gives rise to a complex line field Λ on U which is everywhere transverse to $T^{1,0}(M_t)$. Equation (ii) states that $\xi \in \Lambda$ and (i) determines a unique vector in Λ. □

Remark. In fact c is real-valued because $c = \partial\bar{\partial}\rho(\xi, \bar{\xi})$.

We define $\widetilde{N} = \frac{1}{2}(\xi + \bar{\xi})$ and $\widetilde{T} = J\widetilde{N} = \frac{i}{2}(\xi - \bar{\xi})$. Then $d\rho(\widetilde{T}) = 0$ whence $\widetilde{T}$ is tangent to the level sets of ρ. We let T be the restriction of $\widetilde{T}$ to M.

Lemma 3.2. *The vector field T is the unique solution to the equations*

(i) $\theta(T) = 1$.
(ii) $\iota_T d\theta = 0$.

Proof. The lemma follows immediately from the observation that $2i\partial\bar{\partial}\rho|T(M) \otimes \mathbb{C} = d\theta$ by taking real parts in (ii) of Lemma 3.1. □

Let F be the complementary vector bundle to H in $T(M)$ generated by T. We define $E \subset T(M) \otimes \mathbb{C}$ by

$$E = T^{1,0}(M) \oplus (F \otimes \mathbb{C}).$$

We next observe (following [K2]) that there is a natural isomorphism of complex vector bundles

$$\tau : T^{1,0}(U)|M \longrightarrow E.$$

Indeed the map inverse to τ is induced by the composition

$$T(M) \otimes \mathbb{C} \longrightarrow T(U) \otimes \mathbb{C}|M \longrightarrow T^{1,0}(U)|M$$

which is a surjection with kernel $T^{0,1}(M)$. We observe that $\tau(\xi) = -2iT$ and $\tau|T^{1,0}(M) = id$.

We recall that we can form a complex $(\mathcal{A}^{0,\cdot}(M), \bar\partial_b)$ where

$$\mathcal{A}^{0,i}(M) = \Gamma\left(M, \Lambda^i T^{0,1}(M)^*\right)$$

and $\bar\partial_b$ is defined as follows. Given $\varphi \in \mathcal{A}^{0,i}(M)$ extend it to $\tilde\varphi \in \mathcal{A}^{0,i}(U)$. Then $\bar\partial_b\varphi$ is defined to be the restriction of $\bar\partial\,\tilde\varphi$ to M. The integrability of $T^{0,1}(M)$ implies that $\bar\partial_b$ is well-defined. We say $f \in \mathcal{A}^{0,0}(M)$ is a CR-function if $\bar\partial_b f = 0$. We say that a complex vector bundle over M is CR if it has an atlas with CR-transition functions. If E' is such a bundle then we may twist the above complex by E' to form a complex $(\mathcal{A}^{0,\cdot}(M, E'), \bar\partial_b)$. If E' is the restriction to M of a holomorphic bundle E'' on U then E' is CR and the above complex may be obtained as the boundary complex (i.e. with boundary operator defined using extensions to U as above) associated to the twisted Dolbeault complex $(\mathcal{A}^{0,\cdot}(U, E''), \bar\partial)$ as in the case of trivial coefficients above. Thus we obtain a complex $\left(\mathcal{A}^{0,\cdot}\left(M, T^{1,0}(U)|M\right), \bar\partial_b\right)$ and by transport of structure via τ a complex $(\mathcal{A}^{0,\cdot}(M, E), \bar\partial_b)$. We will abbreviate this complex to $(\mathcal{K}^\cdot(M), \bar\partial_b)$. Thus

$$\mathcal{K}^i(M) = \Gamma\left(M, \left(\Lambda^i T^{1,0}(M)^*\right) \otimes E\right).$$

We now give a formula for $\bar\partial_b$ following [Ak1]. Let $P : T(M) \otimes \mathbb{C} \longrightarrow E$ be the projection with kernel $T^{0,1}(M)$, $\omega \in \mathcal{A}^{0,q}(M, E)$ and $Z_1, \ldots, Z_{q+1}$ be C^∞–sections of $T^{1,0}(M)$. Then

$$\begin{aligned}\bar\partial_b\omega(\bar Z_1, \ldots, \bar Z_{q+1}) &= \sum_{i=1}^{q+1}(-1)^{i+1}P\left([\bar Z_i, \omega(\bar Z_1, \ldots, \hat{\bar Z}_i, \ldots, \bar Z_{q+1})]\right) + \\ &\quad \sum_{1\le i<j\le q+1}^{q+1} (-1)^{i+j}\omega\left([\bar Z_i, \bar Z_j], \bar Z_1, \ldots, \hat{\bar Z}_i, \ldots, \hat{\bar Z}_j, \ldots, \bar Z_{q+1}\right).\end{aligned}$$

We extend τ to a map $\tau : \mathcal{A}^{0,q}\left(U, T^{1,0}(U)\right) \longrightarrow \mathcal{A}^{0,q}(M, E)$ by

$$\tau\varphi(\bar Z_1, \ldots, \bar Z_q) = \tau\left(\varphi(\bar Z_1, \ldots, \bar Z_q)\right)$$

and $\bar Z_1, \ldots, \bar Z_q$ as above. We can now check that Akahori's formula for $\bar\partial_b$ agrees with that induced by the $\bar\partial$–operator on U as described above. It suffices to prove the following lemma.

Lemma 3.3. *Let $\mu \in \mathcal{A}^{0,q}\left(U, T^{1,0}(U)\right)$. Then*

$$\tau\bar\partial\mu = \bar\partial_b\tau\mu.$$

Proof. Let $Z_1, Z_2, \ldots, Z_{q+1}$ be as above. It suffices to prove that both sides of the above formula agree on $\bar Z_1, \ldots, \bar Z_{q+1}$. We extend the Z_i's to sections of $T^{1,0}(U)$ in a neighbourhood of M such that $\partial\rho(Z_i) = 0$, $1 \le i \le q+1$, on M. We extend τ to a homomorphism $\tilde\tau : T^{1,0}(U) \longrightarrow \ker d\rho$ (near M) by $\tilde\tau = id - \partial\rho \otimes \xi$. By a standard formula, see formula 4.4 of [FN2], letting π denote the projection from

$T(U) \otimes \mathbb{C}$ onto $T^{1,0}(U)$ along $T^{0,1}(U)$ we have

$$\bar{\partial}\mu(\bar{Z}_1,\ldots,\bar{Z}_{q+1}) = \sum_{i=1}^{q+1}(-1)^{i+1}\pi\left([\bar{Z}_i,\mu(\bar{Z}_1,\ldots,\hat{\bar{Z}}_i,\ldots,\bar{Z}_{q+1})]\right) + \sum_{1\le i<j\le q+1}^{q+1}(-1)^{i+j}\mu\left([\bar{Z}_i,\bar{Z}_j],\ldots,\hat{\bar{Z}}_i,\ldots,\hat{\bar{Z}}_j,\ldots,\bar{Z}_{q+1}\right).$$

Now $\mu - \tilde{\tau}\mu$ takes values in $T^{0,1}(U)$ by definition and consequently

$$\pi\left([\bar{Z}_i,(\mu-\tilde{\tau}\mu)(\bar{Z}_1,\ldots,\hat{\bar{Z}}_i,\ldots,\bar{Z}_{q+1})]\right) = 0.$$

We obtain

$$\tau\bar{\partial}\mu(\bar{Z}_1,\ldots,\bar{Z}_{q+1}) = \sum_{i=1}^{q+1}(-1)^{i+1}\tau\pi\left([\bar{Z}_i,\tilde{\tau}\mu(\bar{Z}_1,\ldots,\hat{\bar{Z}}_i,\ldots,\bar{Z}_{q+1})]\right) + \sum_{1\le i<j\le q+1}^{q+1}(-1)^{i+j}\tau\mu\left([\bar{Z}_i,\bar{Z}_j],\ldots,\hat{\bar{Z}}_i,\ldots,\hat{\bar{Z}}_j,\ldots,\bar{Z}_{q+1}\right).$$

Now the complex vector fields $\bar{Z}_i$ and $\tilde{\tau}\mu(\bar{Z}_1,\ldots,\hat{\bar{Z}}_i,\ldots,\bar{Z}_{q+1})$ on U take values on M which are in $T(M)\otimes\mathbb{C}$. Hence their bracket depends only on their values on M and we may replace $\tilde{\tau}\mu$ by $\tau\mu$ in the above expression. Since the value of the bracket is in $T(M)\otimes\mathbb{C}$, we are done if we can prove $\tau\circ\pi|T(M)\otimes\mathbb{C} = P$. Both sides annihilate $T^{0,1}(M)$ and are the identity on $T^{1,0}(M)$. Thus it suffices to prove they agree on the vector field T. Since $T \in \Gamma(M,E)$ we have $P(T) = T$. Also $\tau\circ\pi(T) = \tau\left(\frac{i}{2}\xi\right) = \frac{i}{2}\tau(\xi) = T$. □

In what follows we will need the analogue of Lemma 3.3 for scalar-valued $(0,q)$-forms. Let $j : M \longrightarrow U$ be the inclusion. We define the restriction $j^* : \mathcal{A}^{0,q}(U) \longrightarrow \mathcal{A}^{0,q}(M)$, $q \ge 0$, by

$$j^*\mu(\bar{Z}_1,\ldots,\bar{Z}_q) = \mu\left(dj(\bar{Z}_1),\ldots,dj(\bar{Z}_{q+1})\right).$$

We have defined $\bar{\partial}_b$ above such that if $\mu \in \mathcal{A}^{0,q}(U)$ then

$$j^*\bar{\partial}\mu = \bar{\partial}_b j^*\mu.$$

Lemma 3.4. *Let $\mu \in \mathcal{A}^{0,q}(M)$. Then*

$$\bar{\partial}_b\mu(\bar{Z}_1,\ldots,\bar{Z}_{q+1}) = \sum_{i=1}^{q+1}(-1)^{i-1}\bar{Z}_i\mu(\bar{Z}_1,\ldots,\hat{\bar{Z}}_i,\ldots,\bar{Z}_q) + \sum_{1\le i<j\le q+1}^{i+1}\mu\left([\bar{Z}_i,\bar{Z}_j],\ldots,\hat{\bar{Z}}_i,\ldots,\hat{\bar{Z}}_j,\ldots,\bar{Z}_{q+1}\right).$$

We have obtained maps of complexes $j^* : \mathcal{A}^{0,\cdot}(U) \longrightarrow \mathcal{A}^{0,\cdot}(M)$ and $\tau : \mathcal{L}^\cdot(U) \longrightarrow \mathcal{K}^\cdot(M)$. We will often abbreviate $\mathcal{K}^\cdot(M)$ to $\mathcal{K}$. The next theorem is of central importance to us. The proof is contained in pages 81 and 82 of [Y]. We leave the reader to check that the argument there applies when the coefficients Ω^p in [Y] are replaced by $T^{1,0}(U)$.

Theorem 3.1. *The map τ induces isomorphisms on cohomology groups of degree i, $1 \le i \le n-2$.*

Corollary.

(i) *suppose* $\dim V = 3$. *Then τ induces an isomorphism on first cohomology.*
(ii) *Suppose* $\dim V \ge 4$. *Then τ induces an isomorphism on first and second cohomology.*

We next explain the connection between Kuranishi's complex $(\mathcal{K}^{\cdot}(M), \bar\partial_b)$ and deformation theory. If (H', J') is a CR-structure near (H, J) then its $(0,1)$–distribution may be represented as the graph of a bundle map $\varphi : T^{1,0}(M) \longrightarrow E$. The form φ will satisfy the integrability condition [Ak1]

$$S(\varphi) := \bar\partial_b\varphi + \tfrac{1}{2}R_2(\varphi) + R_3(\varphi) = 0$$

where $R_2(\varphi)$ and $R_3(\varphi)$ are elements of $\mathcal{K}^2$ given by the formulas

$$\begin{aligned} R_2(\varphi)(\bar Z, \overline{W}) &= P\left([\varphi(\bar Z), \varphi(\overline{W})]\right) - P\left([\varphi(\overline{W}), \varphi(\bar Z)]\right) + \\ &\quad 2\left[\varphi(Q([\varphi(\bar Z), \overline{W}])) - \varphi(Q([\varphi(\overline{W}), \bar Z]))\right] \end{aligned}$$

and

$$R_3(\varphi)(\bar Z, \overline{W}) = -\varphi\left(Q([\varphi(\bar Z), \varphi(\overline{W})])\right).$$

Here $Q : T(M) \otimes \mathbb{C} \longrightarrow T^{0,1}(M)$ denotes the projection with kernel E.

Remarks. It is important to observe that by defining $\mathcal{K}^0(M) = \Gamma(M, E) \cong \Gamma\left(M, T^{1,0}(U)|M\right)$ we are dividing out by the isotopies of M in U; hence, we are taking care of the dependence of M on the choice of sphere S_ε used to form the link. Kuranishi refers to such isotopies as "wiggles" in [K2].

We will not describe Kuranishi's deformation functor. The reader will find a description of this functor in [Ak2], §6.

We now describe the (formal) parameter space $\operatorname{Spec} R_{\mathcal{K}}$ for Kuranishi's CR-deformation theory. Choose a Hermitian metric on E and let $\bar\partial_b^*$ be the resulting formal adjoint to $\bar\partial_b$ on $\mathcal{K}^{\cdot}(M)$. We define $C^1 \subset \mathcal{K}^1(M)$ by $C^1 = \ker\left(\partial_b^* : \mathcal{K}^1(M) \longrightarrow \mathcal{K}^0(M)\right)$. Now let $A \in \operatorname{Obj} \mathcal{A}$ with maximal ideal $\mathfrak{m}$. We define a functor $\mathcal{K}_M : \mathcal{A} \longrightarrow$ **Sets** by

$$\mathcal{K}_M(A) = \{\varphi \in C^1 \otimes \mathfrak{m} : S(\varphi) = 0\}.$$

Theorem 3.2. *The functor $\mathcal{K}_M$ is pro-representable by a complete local Noetherian $\mathbb{C}$-algebra $R_{\mathcal{K}}$ provided* $\dim M \ge 5$.

Proof. If $\dim M \ge 5$ then $\dim H^1\left(\mathcal{K}^{\cdot}(M)\right) < \infty$ by [Ta], Theorem 6.5. Theorem 3.2 then follows from Theorem 1.2. □

4. The Global Tangent Complex of a Complex Analytic Space

In this section we will describe Palamodov's construction [P] of the (global) tangent complex L_X associated to a complex analytic space and we will give a new construction of the spectral sequence of tangent cohomology. In this chapter we will make extensive use of [F].

A simplicial scheme of ringed spaces W_* consists of a simplicial complex A, a family $\{W_\alpha\}_{\alpha \in A}$ of ringed spaces and morphisms $p_{\alpha\beta} : W_\beta \longrightarrow W_\alpha$ for every pair of simplices α, β with $\alpha \subset \beta$ such that $p_{\alpha\beta}$ satisfy the usual compatibility conditions. We may express these conditions precisely by saying that W_* is a contravariant functor from A, considered as a category in the usual way (the objects are the simplices, the morphisms inclusions of faces), to the category of ringed spaces. We let $\mathcal{O}_{W_*}$ be the structure sheaf of W_*. Thus for every simplex $\alpha \in A$ we have the sheaf $\mathcal{O}_{W_\alpha}$ over W_α and for every pair $\alpha, \beta \subset A$ with $\alpha \subset \beta$ we have a structure map

$$r_{\beta\alpha} : p_{\alpha\beta}^{-1}\mathcal{O}_{W_\alpha} \longrightarrow \mathcal{O}_{W_\beta}.$$

An $\mathcal{O}_{W_*}$-module M_* is a family $\{M_\alpha\}_{\alpha \in A}$ of $\mathcal{O}_{W_\alpha}$-modules M_α together with compatible structure maps

$$\mu_{\beta\alpha} : p_{\alpha\beta}^* M_\alpha \longrightarrow M_\beta$$

for any pair α, β with $\alpha \subset \beta$. We say M_* is *coherent (resp. locally-free)* if all the M_α are coherent (resp. free of finite rank). We say M_* is *locally acyclic* if M_α is acyclic for all $\alpha \in A$. We let $\mathrm{Mod}(\mathcal{O}_{W_*})$ (resp. $\mathrm{Coh}(\mathcal{O}_{W_*})$) be the category of $\mathcal{O}_{W_*}$-modules (resp. coherent $\mathcal{O}_{W_*}$-modules). A graded commutative $\mathcal{O}_{W_*}$-algebra S_* is an $\mathcal{O}_{W_*}$-module such that S_{W_α} is a graded commutative $\mathcal{O}_{W_\alpha}$-algebra for all $\alpha \in A$ and the structure maps are graded algebra homomorphisms. If S_* is such an algebra, then we may form the graded S_*-module $\Omega^1_{S_*}$ of differentials. This module is characterized by the following universal property. There is a derivation of degree zero $d : S_* \longrightarrow \Omega^1_{S_*}$ such that for any graded S_*-module M_*, d induces an isomorphism

$$d^* : \mathrm{Hom}_{S_*}\left(\Omega^1_{S_*}, M_*\right) \longrightarrow \mathrm{Der}(S_*, M_*).$$

Here $d^*(T) = T \circ d$. In the above we interpret Hom to mean the *direct sum* of the homomorphisms of degree n as n varies and Der to denote the corresponding direct sum of the derivations of degree n as n varies. In case $\mathcal{O}_{W_*}$ is a sheaf of $\mathbb{C}$-algebras $\mathrm{Der}(S_*, M)$ will denote the derivations annihilating $\mathbb{C}$.

Let $W_* = \{W_\alpha\}_{\alpha \in A}$ and $V_* = \{V_\beta\}_{\beta \in B}$ be simplicial ringed spaces. Then a morphism $f : W_* \longrightarrow V_*$ consists of a simplicial map $\tau : A \longrightarrow B$ and a compatible family of morphisms $\{f_\alpha\}_{\alpha \in A}$ with

$$f_\alpha : W_\alpha \longrightarrow V_{\tau(\alpha)}.$$

Given such a morphism f there are functors

$$\begin{aligned} f^* &: \mathrm{Mod}(\mathcal{O}_{V_*}) \longrightarrow \mathrm{Mod}(\mathcal{O}_{W_*}) \\ f_* &: \mathrm{Mod}(\mathcal{O}_{W_*}) \longrightarrow \mathrm{Mod}(\mathcal{O}_{W_*}) \end{aligned}$$

defined as follows, [F], page 28. We have

$$f^*(N_*)_\alpha = f_\alpha^*\left(N_{\tau(\alpha)}\right).$$

The definition of f_* is more complicated and will occupy the next paragraph.

We begin by recalling that if $\beta \in B$ the star of β, $\mathrm{St}(\beta)$, is defined by

$$\mathrm{St}(\beta) = \{\beta' \in B : \beta \subset \beta'\}.$$

For $\beta \in B$ we define $A_\beta \subset A$ by $A_\beta = \tau^{-1}(\mathrm{St}(\beta))$. Thus $\alpha \in A_\beta$ if and only if $\tau(\alpha) \supset \beta$. Suppose now that $\alpha \in A_\beta$. We let $f_{\beta\alpha}$ be the composition of $f_\alpha : W_\alpha \longrightarrow V_{\tau(\alpha)}$ with the structure map $V_{\tau(\alpha)} \longrightarrow V_\beta$. We define C_β by $C_\beta = \{(\alpha,\gamma) : \alpha,\gamma \in A_\beta, \alpha \subset \gamma\}$. Finally if $(\alpha,\gamma) \in C_\beta$ we define $\varphi_\beta : (f_{\beta\alpha})_* M_\alpha \longrightarrow (f_{\beta\gamma})_* M_\gamma$ as follows. Observe that if $\alpha \subset \gamma$ we have a structure map $\mu_{\gamma\alpha} : M_\alpha \longrightarrow (p_{\alpha\gamma})_* M_\gamma$ whence a map $(f_{\beta\alpha})_*(\mu_{\gamma\alpha}) : (f_{\beta\alpha})_* M_\alpha \longrightarrow (f_{\beta\gamma})_* M_\gamma$ (note that $f_{\beta\alpha} \circ p_{\alpha\gamma} = f_{\beta\gamma}$). We define $\varphi_\beta = (f_{\beta\alpha})_*(\mu_{\gamma\alpha})$. We then define

$$d_\beta : \prod_{\alpha \in A_\beta} (f_{\beta\alpha})_* M_\alpha \longrightarrow \prod_{(\alpha,\gamma)\in C_\beta} (f_{\beta\gamma})_* M_\gamma$$

by

$$d_\beta(x)_{(\alpha,\gamma)} = \varphi_\beta(x_\alpha) - x_\gamma.$$

Finally we define

$$f_*(M_*)_\beta = \ker d_\beta.$$

Given $\alpha \in A$ we define a functor $p_\alpha^* : \mathrm{Mod}(\mathcal{O}_{W_\alpha}) \longrightarrow \mathrm{Mod}(\mathcal{O}_{W_*})$ by

$$(p_\alpha^* M_\alpha)_\beta = \begin{cases} p_{\alpha\beta}^* M_\alpha & \alpha \subseteq \beta. \\ 0, & \text{otherwise.} \end{cases}$$

We also define a functor $\mathrm{res}_\alpha : \mathrm{Mod}(\mathcal{O}_{W_*}) \longrightarrow \mathrm{Mod}(\mathcal{O}_{W_\alpha})$ by

$$\mathrm{res}_\alpha(M_*) = M_\alpha.$$

Remark. The functors p_α^* and res_α are defined on morphisms in the obvious way. In particular the structure map (for $\alpha \subset \beta$) $\mu_{\beta\alpha} : p_{\alpha\beta}^*(p_\alpha^* M_\alpha)_\alpha = p_{\alpha\beta}^* M_\alpha \longrightarrow (p_\alpha^* M_\alpha)_\beta = p_{\alpha\beta}^* M_\alpha$ is the identity map.

Lemma 4.1. *The functor p_α^* is left adjoint to res_α, that is*

$$\mathrm{Hom}_{\mathcal{O}_{W_*}}(p_\alpha^* M_\alpha, N_*) = \mathrm{Hom}_{\mathcal{O}_{W_\alpha}}(M_\alpha, N_\alpha).$$

Proof. Given an element φ_* on the left-hand side then φ_α is an element of the right-hand side because

$$(p_\alpha^* M_\alpha)_\alpha = M_\alpha.$$

Now let $\varphi_a \in \mathrm{Hom}_{\mathcal{O}_{W_\alpha}}(M_\alpha, N_\alpha)$ be given. We will show that φ_α determines a unique $\varphi_* \in \mathrm{Hom}_{\mathcal{O}_{W_*}}(M_*, N_*)$ with $(\varphi_*)_\alpha = \varphi_\alpha$. Clearly $\varphi_\beta = 0$ unless $\alpha \subset \beta$. If φ_* is a morphism, we have a diagram

$$\begin{array}{ccc} p_{\alpha\beta}^* M_\alpha & \xrightarrow{p_{\alpha\beta}^*\varphi_\alpha} & p_{\alpha\beta}^* N_\alpha \\ \Big\downarrow{\scriptstyle id} & & \Big\downarrow \\ (p_\alpha^* M_\alpha)_\beta & \xrightarrow{\varphi_\beta} & N_\beta \end{array}$$

Thus φ_α determines φ_β for $\beta \supset \alpha$. □

We now give an important definition form [F], page 33.

Definition. An $\mathcal{O}_{W_*}$–module F_* is *free* if it is a direct sum of $p_\alpha^* F_\alpha$ as α varies with F_α a free $\mathcal{O}_{W_\alpha}$–module of finite rank. If A_* is an $\mathcal{O}_{W_*}$–algebra then an A_*–module M is *free* if it is of the form $F_* \otimes_{\mathcal{O}_{W_*}} A_*$ where F_* is a free $\mathcal{O}_{W_*}$–module.

Remark. The idea behind this definition is the following. Suppose $\alpha = (i_0, i_1)$ is a one-simplex of A. If F_* is free then there exists a free $\mathcal{O}_{W_\alpha}$–module F' (the space of "new generators") such that

$$F_\alpha = p_{i_0\alpha}^* F_{i_0} \oplus p_{i_1\alpha}^* F_{i_1} \oplus F'.$$

Note (by considering ranks) that $\mathcal{O}_{W_*}$ is not free unless $\dim A = 0$. We observe that in the case that W_* is a Stein complex-analytic simplicial ringed space then the free $\mathcal{O}_{W_*}$–modules are the projective objects in $\mathrm{Coh}(\mathcal{O}_{W_*})$, see [F], page 30.

A graded $\mathcal{O}_{W_*}$–algebra B_* is called *free* if it is isomorphic to the (graded) symmetric algebra $S(F_*^\cdot)$ where $F_*^\cdot$ is a free graded module with $F_*^i = 0$ for $i \geq 0$. Note that B_* free implies $B_*^0 = \mathcal{O}_{W_*}$. A differential graded algebra is free if its underlying graded algebra is free.

Now let B_* be a graded $\mathcal{O}_{W_*}$–algebra. A resolvent of B_* is a free differential graded $\mathcal{O}_{W_*}$–algebra R_* together with a surjective quasi-isomorphism $\varepsilon : R_* \longrightarrow B_*$. We will often use the notation $\bar{R}_*$ for B_* and $\bar{R}_\alpha$ for B_α in what follows.

We now give the two most important definitions of this chapter, they are taken from [F].

Definition. Let X be a complex analytic space. Then a resolvent $\mathfrak{g}$ for X consists of

(i) A locally finite cover $\mathcal{K} = (X_i)_{i\in I}$ of X by Stein compacts (we let X_* be the resulting simplicial scheme of Stein compacts and $\mathcal{N}$ be the nerve of $\mathcal{K}$).
(ii) A closed embedding $i : X_* \longrightarrow W_*$ where W_* is a smooth simplicial scheme of Stein compacts (with the same underlying simplicial complex $\mathcal{N}$ and $\tau = id_{\mathcal{N}}$).
(iii) A resolvent R_* of $i_*\mathcal{O}_{X_*}$ over $\mathcal{O}_{W_*}$ (we consider $i_*\mathcal{O}_{X_*}$ to be a graded $\mathcal{O}_{W_*}$–module with all graded pieces of non-zero degree equal to zero).

Remarks. We define a simplicial scheme of Stein compacts to be smooth if $\Omega^1_{\mathcal{O}_{W_*}} = \Omega^1_{W_*}$ is a free $\mathcal{O}_{W_*}$–module (here and in what follows free is to be interpreted in terms of the above definition). This is a considerably stronger requirement than saying that the pieces W_α, $\alpha \in \mathcal{N}$, are smooth. In practice we satisfy this condition by choosing closed embeddings $\varphi_i : X_i \longrightarrow W_i$, $i \in I$, with W_i a smooth Stein compact then defining W_α to be the product $W_{i_0} \times W_{i_1} \times \cdots \times W_{i_p}$ where $i_0, i_1, \ldots, i_p$ are the vertices of α and $p_{\alpha\beta}$ to be the projection map on the factors corresponding to α. Then $\Omega^1_{W_*} = \bigoplus_{i\in I} p_i^*\Omega^1_{W_i}$ and consequently $\Omega^1_{W_*}$ is free. We define $X_\alpha = X_{i_0} \cap \cdots \cap X_{i_p}$ and i_α to be the composition of $\varphi_{i_0} \times \cdots \times \varphi_{i_p}$ with the diagonal $X_\alpha \longrightarrow X_{i_0} \times \cdots \times X_{i_p}$. Note also that we will assume that there exists an open Stein covering $\mathcal{V} = \{V_i\}_{i\in I}$ with $X_i \subset V_i$, $i \in I$, together with Stein domains $\widetilde{W}_i$ with $W_i \subset \widetilde{W}_i$ such that $\varphi_i : X_i \longrightarrow W_i$ extends to an embedding $\tilde{\varphi}_i : V_i \longrightarrow \widetilde{W}_i$. By [Bi], Chapter IV, Lemma 3.7, we may assume $\text{Nerve}(\mathcal{V}) = \text{Nerve}(\mathcal{K})$. The $\tilde{\varphi}_i$'s then fit together as above to give an embedding $\tilde{i} : V_* \longrightarrow \widetilde{W}_*$. Here $\widetilde{W}_\alpha = \widetilde{W}_{i_0} \times \cdots \times \widetilde{W}_{i_p}$ where α is as above. Also if U_i denotes the interior of K_i, $i \in I$, we will assume that $\mathcal{U} = \{U_i\}_{i\in I}$ covers X. By [Bi], loc. cit., we may again assume that $\text{Nerve}(\mathcal{U}) = \mathcal{N}$. For details of the construction of W_* and i we refer to [F], Lemma 2.10, or [P], page 159. We will often drop the subscript $*$ on simplicial modules and algebras, in particular we will often write R instead of R_* for the above resolvent. If $\alpha \in \mathcal{N}$ we will use $|\alpha|$ to denote the dimension of the simplex α.

We now give more details concerning the simplicial scheme of Stein compacts X_*. If $\alpha \in \mathcal{N}$ we give $X_\alpha = \bigcap_{i\in\alpha} X_i$ the structure sheaf $\mathcal{O}_X \mid X_\alpha$. If $\alpha, \beta \in X$ with $\alpha \subset \beta$ we take $p_{\alpha\beta} = j_{\alpha\beta}$ where $j_{\alpha\beta} : X_\beta \longrightarrow X_\alpha$ is the inclusion. We note that for α, β as above we have a commutative diagram

$$\begin{array}{ccc} W_\beta & \xrightarrow{p_{\alpha\beta}} & W_\alpha \\ \uparrow{\scriptstyle i_\beta} & & \uparrow{\scriptstyle i_\alpha} \\ X_\beta & \xrightarrow{j_{\alpha\beta}} & X_\alpha \end{array}$$

Finally we define the map $j : X_* \longrightarrow X$ of simplicial spaces. Here we consider X to be a simplicial space defined over a one-point simplicial complex γ and $\tau : \mathcal{N} \longrightarrow \gamma$ is the constant map. The inclusions $j_\alpha : X_\alpha \longrightarrow X_{\tau(\alpha)} = X$ then induce the map j of simplicial spaces. We observe that if $M \in \text{Mod}(\mathcal{O}_X)$ then $M_* = j^*M$ has the property that $M_\alpha = M \mid X_\alpha$ and all the structure maps are the identity.

We will abuse notation and write $\mathfrak{g} = (X_*, W_*, R_*)$ omitting explicit reference to the embedding i.

Definition. Let $\mathfrak{g} = (X_*, W_*, R_*)$ and $\mathfrak{g}' = (X'_*, W'_*, R'_*)$ be resolvents for X and X' respectively and $f : X \longrightarrow X'$ be a morphism of complex analytic spaces. Then a morphism $F : \mathfrak{g} \longrightarrow \mathfrak{g}'$ over f consists of morphisms of simplicial schemes of ringed spaces $g : X_* \longrightarrow X'_*$ with $j \circ g = f \circ j$ and $\tilde{f} : W_* \longrightarrow W'_*$ making the following diagram commutative

$$\begin{array}{ccc} W_* & \xrightarrow{\tilde{f}} & W'_* \\ \uparrow{\scriptstyle i} & & \uparrow{\scriptstyle i'} \\ X_* & \xrightarrow{g} & X'_* \end{array}$$

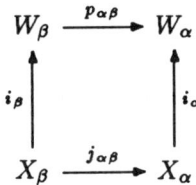

and a morphism of differential graded $\mathcal{O}_{W_*}$–algebras $\tilde{f}^* R'_* \longrightarrow R_*$ making the following diagram commutative

$$\begin{array}{ccc} \tilde{f}^* R'_* & \longrightarrow & R_* \\ \big\downarrow & & \big\downarrow \\ \tilde{f}^* i'_* \mathcal{O}_{X'_*} & \longrightarrow & i_* \mathcal{O}_{X_*} \end{array}$$

where the bottom horizontal arrow is the structure map.

We have the following result [F], Lemma 2.10.

Theorem 4.1.

(i) *Any complex analytic space X has a resolvent $\mathfrak{g}$.*
(ii) *If $\mathfrak{g}$ and $\mathfrak{g}'$ are resolvents for X then there exists a resolvent $\mathfrak{g}''$ for X and morphisms*

$$\mathfrak{g} \longleftarrow \mathfrak{g}'' \longrightarrow \mathfrak{g}'.$$

We now define Palamodov's tangent complex, see [P], page 167.

Definition. Let X be a complex analytic space and $\mathfrak{g}$ be a resolvent for X. Then the tangent complex L_X of X is the differential graded Lie algebra over $\mathbb{C}$ consisting of the direct sum of the compatible graded derivations of R_* of non-negative degree, that is

$$L_X = \mathrm{Der}^+(R_*, R_*).$$

Thus an element of L^i_X, $i \geq 0$, consists of a collection $\{D_\alpha\}_{\alpha \in \mathcal{N}}$ such that D_α is a global derivation over $\mathbb{C}$ of degree i of the sheaf of graded $\mathbb{C}$–algebras $R^\bullet_\alpha$. Moreover if $\alpha, \beta \in \mathcal{N}$ with $\alpha \subset \beta$ we have a commutative diagram

$$\begin{array}{ccc} p^*_{\alpha\beta} R_\alpha & \xrightarrow{r_{\beta\alpha}} & R_\beta \\ {\scriptstyle p^*_{\alpha\beta} D_\alpha}\big\uparrow & & \big\uparrow{\scriptstyle D_\beta} \\ p^*_{\alpha\beta} R_\alpha & \xrightarrow{f} & R_\beta \end{array}$$

Here $r_{\beta\alpha}$ is the structure map.

Remark. We will see at the end of this chapter that up to quasi-isomorphism L_X does not depend on $\mathfrak{g}$.

Lemma 4.2. *$\Omega^1_{R_*}$ is a free R_*–module.*

Proof. We have a split exact sequence

$$0 \longrightarrow \Omega^1_{W_*} \otimes_{\mathcal{O}_{W_*}} R_* \longrightarrow \Omega^1_{R_*} \longrightarrow \Omega^1_{R_*/\mathcal{O}_{W_*}} \longrightarrow 0.$$

By assumption $R_* = S(F^\bullet_*)$ where $F^\bullet_*$ is a free $\mathcal{O}_{W_*}$–module and consequently $\Omega^1_{R_*/\mathcal{O}_{W_*}} = F^\bullet_* \otimes_{\mathcal{O}_{W_*}} R_*$. □

Remark. The reader will verify that if $M^{\bullet}_{*}$ has homogeneous components in $\mathrm{Coh}(\mathcal{O}_{W_*})$ then the natural map $d^* : \mathrm{Hom}_{R_*}(\Omega^1_{R_*}, M^{\bullet}_*) \longrightarrow \mathrm{Der}(R_*, M^{\bullet}_*)$ is an isomorphism.

We will need a basic result, a consequence of the previous lemma and Lemma 7.4. We recall that we abbreviate $i_*\mathcal{O}_{X_*}$ to $\bar{R}_*$.

Lemma 4.3. *The augmentation* $\varepsilon : R_* \longrightarrow \bar{R}_*$ *induces quasi-isomorphisms of complexes* $\varepsilon_* : L_X = \mathrm{Der}^+(R_*, R_*) \longrightarrow \mathrm{Der}^+(R_*, \bar{R}_*)$ *and* $\mathrm{Der}(R_*, R_*) \longrightarrow \mathrm{Der}(R_*, \bar{R}_*)$. □

Corollary. *The inclusion* $\mathrm{Der}^+(R_*, R_*) \longrightarrow \mathrm{Der}(R_*, R_*)$ *is a quasi-isomorphism of differential graded Lie algebras.* □

Remark. We will abbreviate $\mathrm{Der}^+(R_*, \bar{R}_*)$ to $\bar{L}_X$.

We now define the cotangent complex $\mathcal{P}_*$, a complex in $\mathrm{Coh}(\mathcal{O}_{X_*})$, by

$$\mathcal{P}_* = i^* \left(\Omega^1_{R_*} \otimes_{R_*} \mathcal{O}_{W_*}\right).$$

Recall that we are assuming $R^0_* = \mathcal{O}_{W_*}$ whence $\mathcal{O}_{W_*}$ is a quotient of R_*. We observe that if $\alpha \in \mathcal{N}$ then

$$\mathcal{P}_\alpha = i_\alpha^{-1}\Omega^1_{R_\alpha} \otimes_{i_\alpha^{-1}R_\alpha} i_\alpha^{-1}\mathcal{O}_{W_\alpha} \otimes_{i_\alpha^{-1}\mathcal{O}_{W_\alpha}} \mathcal{O}_{X_\alpha}$$

whence

$$\mathcal{P}_\alpha = i_\alpha^{-1}\Omega^1_{R_\alpha} \otimes_{i_\alpha^{-1}R_\alpha} \mathcal{O}_{X_\alpha}.$$

The following lemma is immediate.

Lemma 4.4. *Suppose* $M \in \mathrm{Mod}(\mathcal{O}_{X_*})$. *Then*

$$\mathrm{Der}^+(R_*, i_*M_*) = \mathrm{Hom}_{\mathcal{O}_{X_*}}(\mathcal{P}_*, M_*).$$

The next lemma will allow to calculate the cohomology of the previous complexes.

Lemma 4.5. *The complex* $\mathcal{P}_*$ *is a resolution of* $\Omega^1_{X_*}$ *by free* $\mathcal{O}_{X_*}$*-modules.*

Proof. Since i is an embedding and $\Omega^1_{R_*} \otimes_{R_*} \mathcal{O}_{W_*}$ is a complex of free $\mathcal{O}_{W_*}$-modules it is clear that $\mathcal{P}_*$ is a complex of free $\mathcal{O}_{X_*}$-modules. Let $\alpha \in \mathcal{N}$ and $x \in X_\alpha$. Then it follows from [F], Lemma 1.7, that $\left(i_\alpha^{-1}\Omega^1_{R_\alpha}\right)_x \longrightarrow (\mathcal{P}_\alpha)_x$ is a quasi-isomorphism since $\left(i_\alpha^{-1}R_\alpha\right)_x \longrightarrow \mathcal{O}_{X_\alpha, x}$ is a quasi-isomorphism. Hence $H^i\left((\mathcal{P}_\alpha)_x\right) = 0, i \neq 0$. □

We now consider a general simplicial ringed space $(V_*, \mathcal{O}_{V_*})$ over a simplicial complex $\mathcal{N}$ and construct a family of cochain complexes on $\mathrm{sd}\,\mathcal{N}$, the barycentric subdivision of $\mathcal{N}$, parametrized by an $\mathcal{O}_{V_*}$-algebra S_* and a pair M_*, N_* of S_*-modules. We recall that an n-simplex σ of $\mathrm{sd}\,\mathcal{N}$ is an $n+1$ tuple $\sigma = (\alpha_0, \alpha_1, \ldots, \alpha_n)$ of simplices of $\mathcal{N}$ satisfying $\alpha_0 \subseteq \alpha_1 \subseteq \cdots \subseteq \alpha_n$. Here and in what follows we will let $\alpha_0 = \alpha_0(\sigma)$ and $\alpha_n = \alpha_n(\sigma)$ denote the first and last vertices of σ. We will also let $\mathrm{sd}\,\mathcal{N}^{(n)}$ denote the n-skeleton of $\mathrm{sd}\,\mathcal{N}$.

We recall [Go], §3.3, that a coefficient system $\mathcal{L}$ on a simplicial complex K is a covariant functor $\mathcal{L}$ from K to the category of abelian groups. Given such a functor we may form the complex of (twisted) simplicial cochains $\left(C^{\cdot}(K,\mathcal{L}),d\right)$ where

$$C^n(K,\mathcal{L}) = \prod_{|\sigma|=n} \mathcal{L}(\sigma)$$

and

$$d_n = \sum_{i=0}^{n}(-1)^i d_n^i$$

where $d_n^i = \mathcal{L}(\iota_i)$ with ι_i the inclusion of the ith face of σ into σ.

We now return to S_*, M_*, N_* as above and construct a coefficient system $\operatorname{Hom}_{S_*}(M_*,N_*)$ on $\operatorname{sd}\mathcal{N}$. We will use H to denote this coefficient system if it is not necessary to indicate the particular choice of S_*, M_* and N_* involved. We define H on $\sigma \in \operatorname{sd}\mathcal{N}^{(n)}$ by

$$H(\sigma) = \operatorname{Hom}_{S_{\alpha_0}}\left(M_{\alpha_0}, (p_{\alpha_0,\alpha_n})_* N_{\alpha_n}\right).$$

We observe that $(p_{\alpha_0,\alpha_n})_* N_{\alpha_n}$ is an S_{α_0}-module via the structure map $S_{\alpha_0} \longrightarrow (p_{\alpha_0,\alpha_n})_* S_{\alpha_n}$.

We next define face maps $d_n^i : H(\sigma_i) \longrightarrow H(\sigma)$, $0 \le i \le n$. We define $d_n^i = id$, $1 \le i \le n-1$ (note that in these cases $H(\sigma_i) = H(\sigma)$). We define $d_n^0 : H(\sigma_0) \longrightarrow H(\sigma)$ by observing that by adjunction

$$H(\sigma) = \operatorname{Hom}_{p^*_{\alpha_0,\alpha_1} S_{\alpha_0}}\left(p^*_{\alpha_0,\alpha_1} M_{\alpha_0}, (p_{\alpha_1,\alpha_n})_* N_{\alpha_n}\right).$$

Thus we may define $d_n^0 = \mu^*_{\alpha_1,\alpha_0}$ where $\mu_{\alpha_1,\alpha_0} : p^*_{\alpha_1,\alpha_0} M_{\alpha_0} \longrightarrow M_{\alpha_1}$ is the structure map. Finally to define d_n^n we observe that

$$H(\sigma) = \operatorname{Hom}_{S_{\alpha_0}}\left(M_{\alpha_0}, (p_{\alpha_0,\alpha_{n-1}})_*(p_{\alpha_{n-1},\alpha_n})_* M_{\alpha_n}\right)$$

and we define

$$d_n^n = \left[(p_{\alpha_0,\alpha_{n-1}})_*(\nu_{\alpha_n,\alpha_{n-1}})\right]_*$$

where $\nu_{\alpha_n,\alpha_{n-1}} : N_{\alpha_{n-1}} \longrightarrow (p_{\alpha_{n-1},\alpha_n})_* N_{\alpha_n}$ is the structure map. We may then form the complex of simplicial cochains

$$C^{\cdot}(\operatorname{sd}\mathcal{N},H) = C^{\cdot}\left(\operatorname{sd}\mathcal{N}, \operatorname{Hom}_{S_*}(M_*,N_*)\right).$$

We will especially interested in the special case $V_* = W_*$ as above and $M_* = \Omega^1_{R_*}$ and $S_* = R_*$. In this case we will use $H(\sigma)$ to denote the direct sum of the homomorphisms of non-negative degree (assuming N_* is graded)

$$\begin{aligned} H(\sigma) &= \operatorname{Hom}^+_{R_{\alpha_0}}\left(\Omega^1_{R_{\alpha_0}}, (p_{\alpha_0,\alpha_n})_* N_{\alpha_n}\right) \\ &= \operatorname{Der}^+\left(R_{\alpha_0}, (p_{\alpha_0,\alpha_n})_* N_{\alpha_n}\right). \end{aligned}$$

For this choice of M_* we will write $H = \operatorname{Der}^+(R_*,N_*)$. In this case $C^{\cdot}(\operatorname{sd}\mathcal{N},H^{\cdot})$ is a first quadrant double complex $C^{p,q}$ with $C^{p,q} = C^p(\operatorname{sd}\mathcal{N},H^q)$. We observe that the simplicial cohomology $\check{H}^0\left(C^{\cdot}(\operatorname{sd}\mathcal{N},H)\right)$ is the space $\operatorname{Der}^+(R_*,N_*)$. Thus we have an inclusion

$$\iota : \operatorname{Der}^+(R_*,N_*) \longrightarrow C^{\cdot}\left(\operatorname{sd}\mathcal{N}, \operatorname{Hom}_{R_*}(\Omega^1_{R_*},N_*)\right).$$

In particular if we take $N_* = R_*$ we obtain an inclusion

$$\iota : L_X \longrightarrow C^{\cdot}\left(\operatorname{sd}\mathcal{N}, \operatorname{Hom}_{R_*}(\Omega^1_{R_*}, R_*)\right).$$

Suppose now that $N_* = i_* M_*$ with $M_* \in \operatorname{Mod}(\mathcal{O}_{X_*})$. Let $\alpha, \beta \in \mathcal{N}$ with $\alpha \subset \beta$. Then

$$\begin{aligned}\operatorname{Hom}_{R_\alpha}\left(\Omega^1_{R_\alpha}, (p_{\alpha\beta})_*(i_\beta)_* M_\beta\right) &= \operatorname{Hom}_{R_\alpha}\left(\Omega^1_{R_\alpha}, (i_\alpha)_*(j_{\alpha\beta})_* M_\beta\right) \\ &= \operatorname{Hom}_{\mathcal{O}_{X_\alpha}}\left(\mathcal{P}_\alpha, (j_{\alpha\beta})_* M_\beta\right).\end{aligned}$$

Thus we have an equality of coefficient systems on $\operatorname{sd}\mathcal{N}$

$$\operatorname{Hom}_{R_*}(\Omega^1_{R_*}, i_* M_*) = \operatorname{Hom}_{\mathcal{O}_{X_*}}(\mathcal{P}_*, M_*).$$

The next theorem will allow us to compute the simplicial cohomology groups $\check{H}^i\left(C^{\cdot}\left(\operatorname{sd}\mathcal{N}, \operatorname{Hom}_{\mathcal{O}_{V_*}}(M_*, N_*)\right)\right)$ for many cases of interest. We first need to recall the definition and properties of the functors $\operatorname{Ext}^i_{\mathcal{O}_{V_*}}(\cdot,\cdot)$ from [F], pages 29–33. First by Lemma 2.2 of [F] any $\mathcal{O}_{V_*}$-module N_* admits an injective resolution $I^{\cdot}_*$. It follows that

$$\operatorname{Ext}^i_{\mathcal{O}_{V_*}}(M_*, N_*) = H^i\left(\operatorname{Hom}_{\mathcal{O}_{V_*}}(M_*, I^{\cdot}_*)\right).$$

Now if $M_* \in \operatorname{Coh}(\mathcal{O}_{V_*})$ and $\mathcal{O}_{V_*}$ is coherent then M_* admits a free resolution by Lemma 2.2 of [F]. If M_* is locally free we will construct a free resolution of M_* in Lemma 4.9.

The following lemma is essentially Lemma 2.4 of [F].

Lemma 4.6. *Let M_*, N_* be $\mathcal{O}_{V_*}$-modules with N_* locally acyclic and $F^{\cdot}_*$ be a free resolution of M_*. Then*

$$H^i\left(\operatorname{Hom}_{\mathcal{O}_{V_*}}(F^{\cdot}_*, N_*)\right) = \operatorname{Ext}^i_{\mathcal{O}_{V_*}}(M_*, N_*).$$

Proof. It suffices to prove that $\operatorname{Ext}^i_{\mathcal{O}_{V_*}}(F_*, N_*) = 0$, $i > 0$, for any free $\mathcal{O}_{V_*}$-module F_*. But this follows as in [F], Lemma 2.4. □

Theorem 4.2. *Suppose M_* is locally free and N_* is locally acyclic. Then*

$$\check{H}^i\left(C^{\cdot}\left(\operatorname{sd}\mathcal{N}, \operatorname{Hom}_{\mathcal{O}_{V_*}}(M_*, N_*)\right)\right) = \operatorname{Ext}^i_{\mathcal{O}_{V_*}}(M_*, N_*).$$

Before starting the proof of the theorem, we state and prove a corollary.

Corollary. *Suppose X is a complex analytic space and X_* is the simplicial complex space associated to a Stein cover of X as above. Suppose $M, N \in \operatorname{Mod}(\mathcal{O}_X)$ with M locally free and N either coherent or fine. Put $M_* = j^* M$, $N_* = j^* N$. Then*

$$\check{H}^i\left(C^{\cdot}\left(\operatorname{sd}\mathcal{N}, \operatorname{Hom}_{\mathcal{O}_{X_*}}(M_*, N_*)\right)\right) = H^i\left(X, M^* \otimes_{\mathcal{O}_X} N\right).$$

Proof. By [F], Lemma 2.6 we have

$$\operatorname{Ext}^i_{\mathcal{O}_{X_*}}(j^* M, j^* N) = \operatorname{Ext}^i_{\mathcal{O}_X}(M, N).$$

□

The proof of Theorem 4.2 will follow from the next five lemmas. The first of these follows immediately from the definition of simplicial cochains with values in H.

Lemma 4.7. $C^n(\mathrm{sd}\,\mathcal{N}, H) = \prod_{|\sigma|=n} \mathrm{Hom}_{\mathcal{O}_{V_{\alpha_0}}}(M_{\alpha_0}, (p_{\alpha_0,\alpha_n})_* N_{\alpha_n}).$ □

In order to prove the theorem we will use another description of the above complex as a "standard simplicial resolution", [I], §1.5 and [Go], appendice, associated to an adjoint pair of functors L and R (denoted T and U respectively in [I]). We briefly review the construction.

We recall that if C is a category, the underlying discrete category C^d is the category such that

$$\mathrm{Obj}\ C^d = \mathrm{Obj}\ C$$

and for $a, b \in \mathrm{Obj}\ C^d$ we have

$$\mathrm{Mor}\ (a,b) = \begin{cases} id, & \text{if } a = b. \\ \phi, & \text{otherwise.} \end{cases}$$

Let $\mathcal{A}$ be the category associated to the simplicial complex $\mathcal{N}$ and $\mathcal{A}^d$ the associated discrete category. A simplicial ringed space V_* with underlying category $\mathcal{A}^d$ is then a collection of ringed spaces $\{V_\alpha\}_{\alpha\in\mathcal{N}}$ with structure maps $p_{\alpha\beta}$ only if $\alpha = \beta$ and $p_{\alpha\alpha} = id$. Given a simplicial ringed space V_* on $\mathcal{A}$ we obtain a ringed space V_*^d on $\mathcal{A}^d$ by forgetting the structure maps $p_{\alpha\beta}$, $\alpha \neq \beta$. Thus for V_* as above we have V_*^d and a forgetful functor $R : \mathrm{Mod}(\mathcal{O}_{V_*}) \longrightarrow \mathrm{Mod}(\mathcal{O}_{V_*^d})$ that forgets the structure maps $\mu_{\beta\alpha}$ for $\alpha \neq \beta$.

We also have a functor $L : \mathrm{Mod}(\mathcal{O}_{V_*^d}) \longrightarrow \mathrm{Mod}(\mathcal{O}_{V_*})$ given by

$$L(\{M_\alpha\}_{\alpha\in A}) = \bigoplus_{\alpha\in A} p_\alpha^* M_\alpha.$$

The next lemma follows immediately from Lemma 4.1.

Lemma 4.8. *The functors L and R form an adjoint pair, that is,* $\mathrm{Hom}_{\mathcal{O}_{V_*}}(LM_*, N_*) = \mathrm{Hom}_{\mathcal{O}_{V_*^d}}(M_*, RN_*)$ *for* $M_* \in \mathrm{Mod}(\mathcal{O}_{V_*^d})$, $N_* \in \mathrm{Mod}(\mathcal{O}_{V_*})$. □

Definition. A $\mathcal{O}_{V_*}$-module M_* is said to be left-induced if it satisfies $M_* = LN_*$ for some $N_* \in \mathrm{Mod}(\mathcal{O}_{V_*^d})$.

We note that a free $\mathcal{O}_{V_*}$-module is left-induced by definition; hence, R_* and $\Omega^1_{R_*} \otimes_{R_*} \mathcal{O}_{V_*}$ are left-induced $\mathcal{O}_{V_*}$-modules.

We now consider the corresponding standard resolutions of modules in $\mathrm{Mod}(\mathcal{O}_{V_*})$ associated to the above pair of adjoint functors, see [I], §1.5. By Lemma 4.8, there is a natural transformation $\varepsilon : LR \longrightarrow id$ and corresponding natural transformations $\partial_n^i : (LR)^n \longrightarrow (LR)^{n-1}$, $1 \leq i \leq n$, given by applying ε to the ith factor in the composition on the left. Given $M_* \in \mathrm{Mod}(\mathcal{O}_{V_*})$ we consider the complex $(G.(M), \partial)$ with

$$G_n(M) = (LR)^{n+1}(M), \quad n \geq 0,$$

and

$$\partial_n = \sum_{i=1}^{n+1} (-1)^{i-1} \partial^i_{n+1}.$$

We augment the complex $G.(M)$ by $\varepsilon : G_0(M) \longrightarrow M$. We then have the following results [I], §1.5, [Go], appendices.

Lemma 4.9.

(i) *$(G.(M_*), \partial)$ is a resolution of M_*. If M_* is left-induced the augmented complex is contractible.*

(ii) *If M_* is left-induced the natural map*

$$\varepsilon^* : \mathrm{Hom}_{\mathcal{O}_{V_*}}(M_*, N_*) \longrightarrow \mathrm{Hom}_{\mathcal{O}_{V_*}}(G.(M_*), N_*)$$

is a homotopy equivalence for any $N_* \in \mathrm{Mod}(\mathcal{O}_{V_*})$. □

In our case $G_n(M_*)$ is given by the following formula.

Lemma 4.10. $G_n(M_*)_\beta = \bigoplus_{\alpha_0 \subseteq \cdots \subseteq \alpha_n \subseteq \beta} p^*_{\alpha_0,\beta} M_{\alpha_0}.$

Proof. By induction on n. We have

$$\begin{aligned} G_{k+1}(M_*)_\beta &= LR(G_k(M_*))_\beta \\ &= \left[\bigoplus_{\alpha \in A} p^*_\alpha G_k(M_*)_\alpha \right]_\beta \\ &= \bigoplus_{\alpha \subseteq \beta} p^*_{\alpha\beta} G_k(M_*)_\alpha \\ &= \bigoplus_{\alpha \subseteq \beta} p^*_{\alpha\beta} \bigoplus_{\alpha_0 \subseteq \cdots \subseteq \alpha_k \subseteq \alpha} p^*_{\alpha_0,\alpha} M_{\alpha_0} \\ &= \bigoplus_{\alpha_0 \subseteq \cdots \subseteq \alpha_{k+1} \subseteq \beta} p^*_{\alpha_0,\beta} M_{\alpha_0}. \end{aligned}$$

□

We can now prove our main formula.

Lemma 4.11. *We have a natural isomorphism of cochain complexes*

$$C^{\boldsymbol{\cdot}}\left(\mathrm{sd}\,\mathcal{N}, \mathrm{Hom}_{\mathcal{O}_{V_*}}(M_*, N_*)\right) = \mathrm{Hom}_{\mathcal{O}_{V_*}}(G.(M_*), N_*).$$

Proof.

$$\begin{aligned}
\mathrm{Hom}_{\mathcal{O}_{V_*}}(G_n(M_*), N_*) &= \mathrm{Hom}_{\mathcal{O}_{V_*}}(LRG_{n-1}(M_*), N_*) \\
&= \mathrm{Hom}_{\mathcal{O}_{V_*}}(RG_{n-1}(M_*), RN_*) \\
&= \prod_{\beta \in A} \mathrm{Hom}_{\mathcal{O}_{V_\beta}}(G_{n-1}(M_*)_\beta, N_\beta) \\
&= \prod_{\beta \in A} \prod_{\alpha_0 \subseteq \cdots \subseteq \alpha_{n-1} \subseteq \beta} \mathrm{Hom}_{\mathcal{O}_{V_\beta}}(p^*_{\alpha_0,\beta} M_{\alpha_0}, N_\beta) \\
&= \prod_{\alpha_0 \subseteq \cdots \subseteq \alpha_n} \mathrm{Hom}_{\mathcal{O}_{V_{\alpha_0}}}(M_{\alpha_0}, (p_{\alpha_0,\alpha_n})_* N_{\alpha_n}) \\
&= C^n\left(\mathrm{sd}\,\mathcal{N}, \mathrm{Hom}_{\mathcal{O}_{V_*}}(M_*, N_*)\right).
\end{aligned}$$

The last equality holds by Lemma 4.7.

We leave to the reader the task of checking that the face maps and hence the boundary maps on the two sides coincide. □

We can now prove Theorem 4.2. Indeed if M_* is locally free then $LR(M_*)$ is a free $\mathcal{O}_{V_*}$-module. Thus $G.(M_*)$ is a free resolution of M_* and the theorem follows from Lemma 4.6.

Before stating the next theorem, we recall that by Lemma 4.9 if M_* is left-induced (not necessarily coherent or locally free) then the inclusion of the simplicial cocycles of degree zero

$$\iota : \mathrm{Hom}_{\mathcal{O}_{V_*}}(M_*, N_*) \longrightarrow C^\cdot\left(\mathrm{sd}\,\mathcal{N}, \mathrm{Hom}_{\mathcal{O}_{V_*}}(M_*, N_*)\right)$$

is a homotopy equivalence for any N_*. We obtain a vanishing theorem for the simplicial cohomology groups $\check{H}^i\left(\mathrm{sd}\,\mathcal{N}, \mathrm{Hom}_{\mathcal{O}_{V_*}}(M_*, N_*)\right)$.

Theorem 4.3. *Suppose M_* is a left-induced $\mathcal{O}_{V_*}$-module. Then*

$$\check{H}^i\left(C^\cdot\left(\mathrm{sd}\,\mathcal{N}, \mathrm{Hom}(M_*, N_*)\right)\right) = 0, \quad i > 0.$$

Corollary. *Let (X_*, W_*, R_*) be a resolvent for a complex analytic space X. Then for any $N_* \in \mathrm{Mod}(\mathcal{O}_{W_*})$*

$$\check{H}^i\left(C^\cdot\left(\mathrm{sd}\,\mathcal{N}, \mathrm{Der}^+(R_*, N_*)\right)\right) = 0, \quad i > 0.$$

Proof. $\Omega^1_{R_*} \otimes_{R_*} \mathcal{O}_{W_*}$ is a left-induced $\mathcal{O}_{W_*}$-module. □

Corollary. *Let X be as above, $\mathcal{P}_*$ be the cotangent complex of X and $N_* \in \mathrm{Mod}(\mathcal{O}_{X_*})$. Then*

$$\check{H}^i\left(C^\cdot\left(\mathrm{sd}\,\mathcal{N}, \mathrm{Hom}_{\mathcal{O}_{X_*}}(\mathcal{P}_*, N_*)\right)\right) = 0, \quad i > 0.$$

Proof. Since $i : X_* \longrightarrow W_*$ is an embedding the functor i^* carries left-induced $\mathcal{O}_{W_*}$–modules to left-induced $\mathcal{O}_{X_*}$–modules. Hence $\mathcal{P}_*$ is a left-induced $\mathcal{O}_{X_*}$–module. □

We conclude this chapter by deriving the spectral sequence of tangent cohomology and the independence of the tangent complex L_X on the choice of resolvent $\mathfrak{g}$ (up to quasi-isomorphism of differential graded Lie algebras). We consider the double complex
$C^\bullet(\mathrm{sd}\,\mathcal{N}, \mathrm{Hom}_{\mathcal{O}_{X_*}}(\mathcal{P}_*, \mathcal{O}_{X_*}))$ as above. By Theorem 4.3 the inclusion $\bar{L}_X \longrightarrow C^\bullet(\mathrm{sd}\,\mathcal{N}, H)_{\mathrm{tot}}$ is a quasi-isomorphism. We now investigate the spectral sequence (E_r, d_r) obtained when we filter the above double complex by simplicial degree (i.e. the gradation coming from $\mathrm{sd}\,\mathcal{N}$).

We recall, [P], Proposition 1.9, that there exist coherent sheaves $\mathcal{T}^q$, $q \geq 0$, on X, the tangent cohomology sheaves, with stalks $\mathcal{T}^q_x$ given by

$$\mathcal{T}^q_x = H^q(\mathrm{Der}(R_x, R_x))$$

where R_x is any multiplicative resolution of the analytic local ring $\mathcal{O}_{X,x}$ (see Chapter 5).

Theorem 4.4. *The spectral sequence (E_r, d_r) above converges to $H^\bullet(L_X)$ and has E_2 term given by*

$$E_2^{p,q} = H^p(X, \mathcal{T}^q).$$

The rest of this chapter will be devoted to the proof of this theorem. Since the cohomology of the total complex associated to $C^{p,q}$ coincides with that of $\bar{L}_X$ by Theorem 4.3, the spectral sequence has the correct limit. It remains to compute the E_2–term.

We begin by observing that if N_* is an $\mathcal{O}_{X_*}$–module then
$C^\bullet(\mathrm{sd}\,\mathcal{N}, \mathrm{Hom}(\mathcal{O}_{X_*}, N_*))$ has a very simple description. Indeed it is immediate that

$$C^n(\mathrm{sd}\,\mathcal{N}, \mathrm{Hom}(\mathcal{O}_{X_*}, N_*)) = \prod_{|\sigma|=n} \Gamma(X_{\alpha_n}, N_{\alpha_n}).$$

Also the differentials $\{d_n\}$ are given by somewhat simpler formulas which we now describe.

We have $d_n = \sum_{i=0}^n (-1)^i d_n^i$ with $d_n^i = id$ for $0 \leq i \leq n-1$. Then we define

$$d_n^n : \Gamma\left(X_{\alpha_{n-1}}, N_{\alpha_{n-1}}\right) \longrightarrow \Gamma\left(X_{\alpha_n}, N_{\alpha_n}\right)$$

to be the map on global sections over $X_{\alpha_{n-1}}$ induced by the structure map

$$\mu_{\alpha_n,\alpha_{n-1}} : N_{\alpha_{n-1}} \longrightarrow \left(j_{\alpha_{n-1},\alpha_n}\right)_* N_{\alpha_n}.$$

The next lemma is the key step in the computation of E_2–term.

Lemma 4.12. *Suppose $\alpha, \beta \in \mathcal{N}$, with $\alpha \subset \beta$. Then the structure map $\mu_{\beta\alpha} : j^*_{\alpha\beta}\mathcal{P}_\alpha \longrightarrow \mathcal{P}_\beta$ induces a quasi-isomorphism of complexes of $\mathcal{O}_{X_\beta}$–modules*

$$\mathrm{Hom}_{\mathcal{O}_{X_\beta}}(\mathcal{P}_\beta, \mathcal{O}_{X_\beta}) \longrightarrow \mathrm{Hom}_{\mathcal{O}_{X_\alpha}}\left(\mathcal{P}_\alpha, (j_{\alpha\beta})_*\mathcal{O}_{X_\beta}\right).$$

Proof. Since X_β is Stein it suffices to prove the corresponding statement for sheaves. Let $x \in X_\beta$ and $i \geq 0$ be given. We have a commutative diagram

$$\begin{array}{ccc} \mathrm{Hom}^i_{\mathcal{O}_{X_\beta}}\left(\mathcal{P}_\beta, \mathcal{O}_{X_\beta}\right)_x & \longrightarrow & \mathrm{Hom}^i_{\mathcal{O}_{X_\beta}}\left(j^*_{\alpha\beta}\mathcal{P}_\alpha, \mathcal{O}_{X_\beta}\right)_x \\ \downarrow & & \downarrow \\ \mathrm{Hom}^i_{\mathcal{O}_{X,x}}(\mathcal{P}_{\beta,x}, \mathcal{O}_{X,x}) & \longrightarrow & \mathrm{Hom}^i_{\mathcal{O}_{X,x}}((j^*_{\alpha\beta}\mathcal{P}_\alpha)_x, \mathcal{O}_{X,x}) \end{array}$$

Since $\mathcal{P}_\beta^{-1}$ is coherent the vertical maps are isomorphisms. Thus it suffices to show that the map of complexes induced by the bottom horizontal arrow is a quasi-isomorphism. But $(j^*_{\alpha\beta}\mathcal{P}_\alpha)_x$ and $\mathcal{P}_{\beta,x}$ are both free resolutions for $\Omega^1_{X,x}$ as an $\mathcal{O}_{X,x}$–module and $(\mu_{\beta\alpha})_x$ induces an isomorphism of zero-th cohomology. Hence $(\mu_{\beta\alpha})_x$ is a homotopy equivalence. □

Lemma 4.13. *There is an equality of complexes*

$$E_1 = C^\cdot\left(\mathrm{sd}\,\mathcal{N}, \mathrm{Hom}_{\mathcal{O}_{X_*}}(\mathcal{O}_{X_*}, \mathcal{T}^\cdot)\right).$$

Proof. By definition

$$\begin{aligned} E_1^{k,\ell} &= \prod_{|\sigma|=k} H^\ell\left(\mathrm{Hom}_{\mathcal{O}_{X_{\alpha_0}}}\left(\mathcal{P}_{\alpha_0}, (j_{\alpha_0,\alpha_k})_*\mathcal{O}_{X_{\alpha_k}}\right)\right) \\ &= \prod_{|\sigma|=k} H^\ell\left(\mathrm{Hom}_{\mathcal{O}_{X_{\alpha_k}}}(\mathcal{P}_{\alpha_k}, \mathcal{O}_{X_{\alpha_k}})\right) \\ &= \prod_{|\sigma|=k} \mathcal{T}^\ell(X_{\alpha_k}). \end{aligned}$$

□

The next lemma is an immediate consequence of the corollary to Theorem 4.2 since $\mathcal{T}^q$ is coherent and $\mathcal{O}_X$ is locally free.

Lemma 4.14. $H^p\left(C^\cdot\left(\mathrm{sd}\,\mathcal{N}, \mathrm{Hom}_{\mathcal{O}_{X_*}}(\mathcal{O}_{X_*}, \mathcal{T}^q)\right)\right) = H^p(X, \mathcal{T}^q).$ □

Corollary. $E_2^{p,q} = H^p(X, \mathcal{T}^q).$ □

We can now prove that L_X does not depend on the resolvent $\mathfrak{g}$ of X.

Theorem 4.5. *Suppose* $\mathfrak{g} = (X_*, W_*, R_*)$ *and* $\mathfrak{g}' = (X'_*, W'_*, R'_*)$ *are resolvents for* X *and* L_X *and* L'_X *are the corresponding Palamodov tangent complexes. Then* L_X *and* L'_X *are quasi-isomorphic as differential graded Lie algebras.*

Proof. Let $\mathfrak{g}'' = (X''_*, W''_*, R''_*)$ be a resolvent of X which maps to $\mathfrak{g}$ and $\mathfrak{g}'$ (see Theorem 4.1). We obtain diagrams

$$W_* \xleftarrow{\ \pi\ } W''_* \xrightarrow{\ \pi'\ } W'_*$$

and

$$\begin{array}{ccccc}
L_X & \xrightarrow{\pi} & L''_X & \xleftarrow{\pi'} & L'_X \\
\downarrow & & \downarrow & & \downarrow \\
\bar{L}_X & \xrightarrow{\pi} & \bar{L}''_X & \xleftarrow{\pi'} & \bar{L}'_X
\end{array}$$

where L_X, L'_X and L''_X are the tangent complexes constructed using the resolvents $\mathfrak{g}$, $\mathfrak{g}'$ and $\mathfrak{g}''$ respectively. By Lemma 4.3 it suffices to show that the maps $\bar{L}_X \longrightarrow \bar{L}''_X$ and $\bar{L}'_X \longrightarrow \bar{L}''_X$ are quasi-isomorphisms of complexes (note that the maps $L_X \longrightarrow L''_X$ and $L'_X \longrightarrow L''_X$ are bracket preserving). We now consider the corresponding double complexes as above and obtain a diagram

$$\begin{array}{ccccc}
C^{\bullet}(\mathrm{sd}\,\mathcal{N}, H) & \longrightarrow & C^{\bullet}(\mathrm{sd}\,\mathcal{N}, H'') & \longleftarrow & C^{\bullet}(\mathrm{sd}\,\mathcal{N}, H') \\
\uparrow & & \uparrow & & \uparrow \\
\bar{L}_X & \longrightarrow & \bar{L}''_X & \longleftarrow & \bar{L}'_X
\end{array}$$

By the corollary to Theorem 4.3 the vertical arrows are quasi-isomorphisms. But by Theorem 4.4 the upper horizontal arrows are quasi-isomorphisms because they induce isomorphisms of E_2-terms. (Note that the tangent cohomology sheaves $\mathcal{T}^{\bullet}$ are intrinsically defined and do not depend on the choice of resolvent, [P], Proposition 1.9.) □

5. The Local Tangent Complex Controls the Flat Deformations of an Analytic Local Ring

The material of the chapter is closely related to the preprint [SS1]. In [SS1] the authors parametrize the deformations of a given rational homotopy type X in terms of the gauge equivalence classes of perturbations of the differential of the minimal model of the de Rham algebra $\mathcal{A}^{\bullet}(X)$. As in [SS2], we parametrize the flat deformations of an analytic local $\mathbb{C}$-algebra B by the gauge equivalence classes of perturbations of the differential of a multiplicative (or Tate) resolution R of B.

We begin with some definitions. Let $\boldsymbol{k}$ be a valued field of characteristic zero. We define an analytic local $\boldsymbol{k}$-algebra B to be the quotient of the ring $\boldsymbol{k}\langle z_1, z_2, \ldots, z_m\rangle$ of convergent power series in n variables by an ideal. Given such an algebra B we define a cofibred groupoid $\mathrm{Def}(B; \cdot)$ over the category $\mathcal{A}$ of Artin local $\boldsymbol{k}$-algebras. All the diagrams that follow are in the category of analytic $\boldsymbol{k}$-algebras.

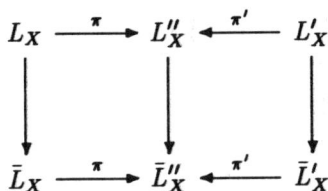

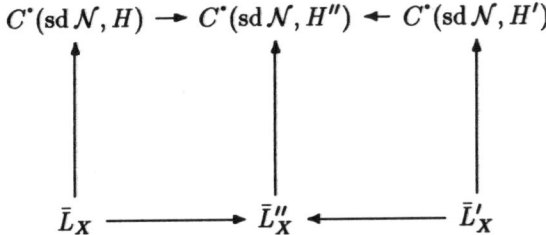

Let $A \in \text{Obj } \mathcal{A}$ and let $\mathfrak{m}$ be the maximal ideal of A. We define the groupoid $\text{Def}(B; A)$ as follows. An object of $\text{Def}(B; A)$ is a flat A–algebra B' and a homomorphism $\rho : B' \longrightarrow B$ of A–algebras such that the following diagram is cocartesian

$$\begin{array}{ccc} B' & \xrightarrow{\quad\rho\quad} & B \\ \uparrow & & \uparrow{\scriptstyle \iota} \\ A & \xrightarrow{\qquad} & \boldsymbol{k} \end{array}$$

We recall that the above diagram is cocartesian if $\rho \otimes \iota : B' \otimes_A \boldsymbol{k} \longrightarrow B$ is an isomorphism. Note that we may identify $B' \otimes_A \boldsymbol{k}$ with $B'/\mathfrak{m}B'$ and consequently ρ induces an isomorphism $\bar{\rho} : B'/\mathfrak{m}B' \longrightarrow B$. We will sometimes refer to ρ as reduction modulo $\mathfrak{m}$.

Remark. If S is an A–module and $s_1, s_2 \in S$ we will write $s_1 \equiv s_2 \bmod \mathfrak{m}$ to indicate that $s_1 - s_2 \in \mathfrak{m}S$. For $s \in S$ we will write $\bar{s}$ for the image of s in $S/\mathfrak{m}S$. We will often abbreviate $S \otimes_A \boldsymbol{k}$ to $\bar{S}$.

Let $B', B'' \in \text{Obj Def}(B; A)$. A morphism $\varphi : B' \longrightarrow B''$ will be an A–algebra homomorphism such that φ induces the identity modulo $\mathfrak{m}$, note $B'/\mathfrak{m}B' = B''/\mathfrak{m}B'' = B$.

By Nakayama's Lemma it follows that the morphisms of $\text{Def}(B; A)$ defined above are isomorphisms. Suppose now that $\varphi : B' \longrightarrow B'$ is an endomorphism in $\text{Def}(B; A)$.

Lemma 5.1. *The endomorphism $\psi = \varphi - id$ of B' is nilpotent.*

Proof. Since $\varphi \equiv id \bmod \mathfrak{m}$ it follows that $\psi(B') \subset \mathfrak{m}B'$ whence $\psi^n(B') \subset \mathfrak{m}^n B'$. But $\mathfrak{m}$ is nilpotent. □

Corollary. *If φ is an endomorphism of* $\text{Def}(B; A)$ *then φ is the exponential of a nilpotent derivation.*

Proof. Put $\delta = \log(\varphi) = \log(id + \psi) = \sum_{i=1}^{\infty} (-1)^{\frac{\psi^i}{i}}$. Then δ is a nilpotent derivation and $\varphi = \exp(\delta)$. □

We next recall the definition of a multiplicative resolution $R^{\bullet}$ of B.

Definition. A differential graded $\boldsymbol{k}$–algebra $(R^{\bullet}, \partial)$ with $R^i = 0$, $i > 0$, together with a surjective homomorphism $\varepsilon : R^{\bullet} \longrightarrow B$ is called a ***multiplicative resolution*** (or ***resolvent***) of B over $\boldsymbol{k}$ if $R^{\bullet}$ is a free graded-commutative $\boldsymbol{k}$–algebra and ε is a quasi-isomorphism.

Remark. We will use the convention that $R_n = R^{-n}$. Thus we will often use the positively graded chain complex $R_\bullet$.

It is standard, [F], Theorem 1.4, that resolvents exists and are unique up to homotopy equivalence.

We now construct another groupoid $\mathrm{Def}(R;\cdot)$ cofibred over $\mathcal{A}$. Let $A \in \mathrm{Obj}\ \mathcal{A}$ with maximal ideal $\mathfrak{m}$. An object of $\mathrm{Def}(R;A)$ is a flat differential graded A–algebra R' and a map of differential graded A–algebras $\rho : R' \longrightarrow R$ such that the following diagram is cocartesian

$$\begin{array}{ccc} R' & \longrightarrow & R \\ \uparrow & & \uparrow \\ A & \longrightarrow & \boldsymbol{k} \end{array}$$

Since R is free it is rigid as a graded commutative algebra and we may assume that $R' \cong R \otimes A$ as algebras ($\otimes$ will denote $\otimes_{\boldsymbol{k}}$). See Lemma 6.5 for a proof of a more general theorem. We will henceforth assume that all deformations R' of R over A have $R \otimes A$ as underlying algebra. However the differential ∂' on R' is not necessarily equal to $\partial \otimes 1$ where ∂ is the differential of R. A morphism φ of the groupoid $\mathrm{Def}(R;A)$ is a homorphism of differential graded algebras such that $\varphi \equiv id \bmod \mathfrak{m}$. Here we identify $R'/\mathfrak{m}R'$ with R. As before we find that every morphism is an isomorphism. We will need a more precise statement. Let $\beta : \mathrm{Hom}_{\boldsymbol{k}}(R,R)\otimes A \longrightarrow \mathrm{Hom}_A(R',R')$ be the canonical isomorphism (here Hom means algebra homomorphisms without regard to the differential and $\beta(\eta\otimes t) = t\big(\eta\otimes(id)\big)$. We observe that β carries $\mathrm{Hom}_{\boldsymbol{k}}(R,R)\otimes\mathfrak{m}$ isomorphically onto $\mathrm{Hom}_A(R',\mathfrak{m}R')$. The proof of the next lemma is analogous to that of Lemma 5.1 (combined with the observation in the line above) and is omitted. We will make the identification $\mathrm{Hom}_{\boldsymbol{k}}(R, R\otimes\mathfrak{m}) = \mathrm{Hom}_{\boldsymbol{k}}(R,R)\otimes\mathfrak{m}$.

Lemma 5.2. *Let φ be a morphism of* $\mathrm{Def}(R;A)$. *Then there exists a derivation δ of degree* 0 *of R with values in $R\otimes\mathfrak{m}$ such that* $\varphi = \exp\big(\beta(\delta)\big)$.

We will need the following lemma.

Lemma 5.3. *Let C' be a flat complex of A–modules and C be the reduction of C' modulo $\mathfrak{m}$. Then $H_q(C) = 0$ implies $H_q(C') = 0$.* □

Proof. By Artinian induction. Let $\mathfrak{J}$ be an ideal of A with $\mathfrak{J}\mathfrak{m} = 0$. For this proof we will use $^{-}$ for reduction modulo $\mathfrak{J}$. By induction we may assume $H_q(\bar{C}') = 0$. By flatness we have a short exact sequence of complexes

$$0 \longrightarrow C' \otimes_A \mathfrak{J} \longrightarrow C' \longrightarrow \bar{C}' \longrightarrow 0$$

and a corresponding long exact sequence of cohomology. But $C' \otimes_A \mathfrak{J} \cong C \otimes_{\boldsymbol{k}} \mathfrak{J}$ whence $H_q(C' \otimes_A \mathfrak{J}) = 0$ and the lemma follows. □

Corollary. *The complex R' is a resolution of $H_0(R')$ by free A–modules.*

Definition. We define the (local) *tangent complex* of the analytic local $\mathbf{k}$–algebra B (or the corresponding analytic germ) to be the differential graded Lie algebra (L, d) such that L is the graded Lie algebra of graded derivations of R of non-negative degree and $d = \operatorname{ad} \partial$. Thus if $\eta \in L^i$ then

$$d\eta = \partial \circ \eta - (-1)^i \eta \circ \partial.$$

Recall that in Chapter 1 we constructed a cofibred groupoid $\mathcal{C}(L; \cdot)$ over the category $\mathcal{A}$. We now compare the groupoids $\mathcal{C}(L; \cdot)$, $\operatorname{Def}(R; \cdot)$ and $\operatorname{Def}(B; \cdot)$.

More precisely let $A \in \operatorname{Obj} \mathcal{A}$ with maximal ideal $\mathfrak{m}$. We define $p : \mathcal{C}(L; A) \longrightarrow \operatorname{Def}(R; A)$ as follows. Let $\eta \in \operatorname{Obj} \mathcal{C}(L; A)$. Then define

$$p(\eta) = (R \otimes A, \beta(\partial) + \beta(\eta)) .$$

Let $\exp(\lambda) \in \operatorname{Mor} \mathcal{C}(L; A)$. Then define $p(\exp(\lambda)) = \beta(\exp(\lambda))$.

We claim p is a natural transformation of cofibred groupoids. It suffices to prove (apply β to this formula)

$$\operatorname{Ad}(\exp(t\lambda))(\partial + \eta) - \partial = \alpha(\exp(t\lambda)) \cdot \eta.$$

But the expression on the left-hand side of this formula gives a possibly different affine action γ of $\exp(L^0 \otimes \mathfrak{m})$ on $L^1 \otimes \mathfrak{m}$. We differentiate with respect to t at $t = 0$ and obtain

$$d\gamma(\lambda) \cdot \eta = [\lambda, \partial + \eta] = [\lambda, \eta] - [\partial, \lambda] = d\alpha(\lambda) \cdot \eta.$$

The claim follows. We now show p is an equivalence.

Lemma 5.4. *The functor p is an equivalence of groupoids.*

Proof. **Surjectivity on isomorphism classes:** Let $R' \in \operatorname{Def}(R; A)$. We have assumed (as algebras) $R' = R \otimes A$. But $\partial' \equiv \beta(\partial) \bmod \mathfrak{m}$ whence $\partial' - \beta(\partial) \in \operatorname{Der}^1_A(R', \mathfrak{m}R')$. Hence there exists $\eta \in L^1 \otimes \mathfrak{m}$ with $\partial' - \beta(\partial) = \beta(\eta)$. But it is immediate that $(\partial')^2 = 0$ implies $d\eta + \eta \circ \eta = d\eta + \frac{1}{2}[\eta, \eta] = 0$. Hence $\eta \in \operatorname{Obj} \mathcal{C}(L; A)$.

Faithful: This is clear.

Full: Let φ be a morphism of $\operatorname{Def}(R; A)$. Then by Lemma 5.2 there exists $\delta \in L^0 \otimes \mathfrak{m}$ such that $\varphi = \exp \beta(\delta) = p(\exp(\delta))$. □

Corollary. *The complete local $\mathbf{k}$–algebra R_L is a pro-representable hull for the functor* Iso $\operatorname{Def}(R; \cdot)$.

Proof. The natural transformation p induces an isomorphism of functors

$$p : \operatorname{Iso} \mathcal{C}(L; \cdot) \longrightarrow \operatorname{Iso} \operatorname{Def}(R; \cdot).$$

But we have seen in Theorem 1.1 that R_L is a pro-representable hull for Iso $\mathcal{C}(L; \cdot)$ for any differential graded Lie algebra L. □

We now define a functor $h : \mathrm{Def}(R;\cdot) \longrightarrow \mathrm{Def}(B;\cdot)$. We define h on objects by

$$h(R') = H_0(R')$$

and on morphisms by

$$h(\varphi) = H_0(\varphi).$$

Here $H_0(\varphi)$ is the morphism induced by φ on the homology of degree 0.

Lemma 5.5. *The functor h is well-defined.*

Proof. The diagram

$$\begin{array}{ccc} R' & \longrightarrow & R \\ \uparrow & & \uparrow \\ A & \longrightarrow & \boldsymbol{k} \end{array} \quad \text{induces} \quad \begin{array}{ccc} H_0(R') & \longrightarrow & B \\ \uparrow & & \uparrow \\ A & \longrightarrow & \boldsymbol{k} \end{array}$$

We must prove that $H_0(R)$ is flat over A and that the right-hand square is cocartesian, that is that the induced map $H_0(R')\otimes_A \boldsymbol{k} \longrightarrow H_0(R'\otimes_A \boldsymbol{k}) = H_0(R) = B$ is an isomorphism. This latter statement follows immediately from the right-exactness of $\otimes_A\boldsymbol{k}$ applied to the short-exact sequence $B_0(R') \longrightarrow C_0(R') \longrightarrow H_0(R')$.

It remains to check that $H_0(R')$ is flat over A. By Lemma 5.3, R' is a resolution of $H_0(R')$ and $R = R'\otimes_A \boldsymbol{k}$. Hence we have $H_i(R) = \mathrm{Tor}_i^A(H_0(R'), \boldsymbol{k})$; $0 \le i < \infty$. But $H_i(R) = 0$, $i > 0$, and consequently $\mathrm{Tor}_1^A(H_0(R'), \boldsymbol{k}) = \{0\}$. It follows from [A2], page 7, that $H_0(R')$ is a flat A-module. □

We now wish to prove that h induces an isomorphism of functors $\mathrm{Iso\ Def}(R;\cdot) \longrightarrow \mathrm{Iso\ Def}(B;\cdot)$. It is not true that h induces an equivalence of groupoids since $h(\varphi)$ does not determine φ (but only the homotopy class of φ) so h is not faithful. However it will suffice to prove that h is full and surjective on isomorphism classes.

We will first consider an intermediate problem. We fix a surjective homomorphism $\varepsilon : \boldsymbol{k}\langle x_1, x_2, \ldots, x_m\rangle \longrightarrow B$. Let $A \in \mathrm{Obj}\ \mathcal{A}$ with maximal ideal $\mathfrak{m}$. We define the groupoid Emb Def(B; A) of embedded deformations of B as follows. An object of Emb Def(B; A) is a deformation $\rho : B' \longrightarrow B$ over A and a surjective homomorphism $\varepsilon' : A\langle x_1, \ldots, x_m\rangle \longrightarrow B'$ such that the following diagram commutes

$$\begin{array}{ccc} A\langle x_1, \ldots, x_m\rangle & \xrightarrow{\rho_0} & \boldsymbol{k}\langle x_1, \ldots, x_m\rangle \\ {\scriptstyle \varepsilon'}\downarrow & & \downarrow{\scriptstyle \varepsilon} \\ B' & \xrightarrow{\rho} & B \end{array}$$

where ρ_0 is the natural map. Let $\varepsilon' : A\langle x_1, \ldots, x_m\rangle \longrightarrow B'$ and $\varepsilon'' : A\langle x_1, \ldots, x_m\rangle \longrightarrow B''$ be objects in Emb Def(B; A). Then a morphism from ε' to ε'' is a morphism

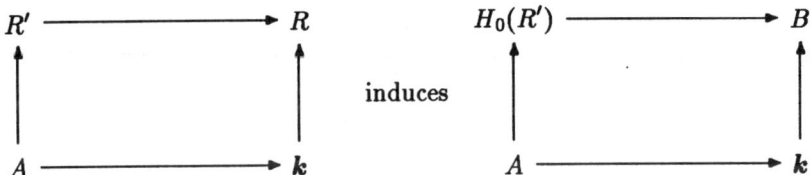

$f : B' \longrightarrow B''$ and a morphism $F : A\langle x_1, \ldots, x_m\rangle \longrightarrow A\langle x_1, \ldots, x_m\rangle$ such that the following diagram commutes

$$\begin{array}{ccc} A\langle x_1,\ldots,x_m\rangle & \xrightarrow{F} & A\langle x_1,\ldots,x_m\rangle \\ \downarrow{\scriptstyle \varepsilon'} & & \downarrow{\scriptstyle \varepsilon''} \\ B' & \xrightarrow{f} & B'' \end{array}$$

Clearly we have a functor

$$\Phi : \text{Emb Def(B; A)} \longrightarrow \text{Def(B; A)}$$

given by $\Phi(\varepsilon' : A\langle x_1, \ldots, x_m\rangle \longrightarrow B') = B'$ and $\Phi((F, f)) = f$.

We will now check that Φ is surjective on isomorphism classes and full. The reader will observe that we do not need the flatness of B'/A for this.

Lemma 5.6. *Φ is surjective on isomorphism classes.*

Proof. Let $B' \in$ Obj $\text{Def}(B; A)$ be given. We have a diagram (with ρ surjective)

$$\begin{array}{ccc} & & B' \\ & \overset{\sigma}{\nearrow} & \downarrow{\scriptstyle \rho} \\ \boldsymbol{k}\langle x_1,\ldots,x_m\rangle & \xrightarrow{\varepsilon} & B \end{array}$$

Since B' is an analytic $\boldsymbol{k}$–algebra, we may lift ε to σ and since B' is an A–algebra we obtain an induced map $\varepsilon' : A\langle x_1, \ldots, x_m\rangle \longrightarrow B'$ such that $\varepsilon'(p \otimes a) = a\sigma(p)$. Clearly $\varepsilon' \equiv \varepsilon \mod \mathfrak{m}$. We claim ε' is surjective. Since $-\otimes_A \boldsymbol{k}$ is right exact we have $\text{cok}(\varepsilon') \otimes_A \boldsymbol{k} = \text{cok}(\varepsilon) = 0$. The claim follows from the strong form of Nakayama's Lemma for modules over Artin local $\boldsymbol{k}$–algebras. □

Lemma 5.7. *The functor Φ is full.*

Proof. Let $f : B' \longrightarrow B''$ be a morphism. We have a lifting problem

$$\begin{array}{ccc} A\langle x_1,\ldots,x_m\rangle & \overset{F}{\dashrightarrow} & A\langle x_1,\ldots,x_m\rangle \\ \downarrow{\scriptstyle \varepsilon'} & & \downarrow{\scriptstyle \varepsilon''} \\ B' & \xrightarrow{f} & B'' \end{array}$$

Suppose we can find $p_i \in A\langle x_1, \ldots, x_m\rangle$ such that $\rho_0(p_i) = x_i$ and $\varepsilon''(p_i) = f(\varepsilon'(x_i))$, $1 \le i \le m$. Let F be the unique algebra homomorphism with $F(x_i) = p_i$, $1 \le i \le m$. Then the above diagram commutes and $\rho_0 \circ F = id \circ \rho_0$. Thus we are reduced to finding p_i, $1 \le i \le m$.

We will now show that the natural map $\varphi : A\langle x_1, \ldots, x_m\rangle \longrightarrow \boldsymbol{k}\langle x_1, \ldots, x_m\rangle \times_B B''$ is onto (here the fibre product is associated to the maps $\varepsilon : \boldsymbol{k}\langle x_1, \ldots, x_m\rangle \longrightarrow B$

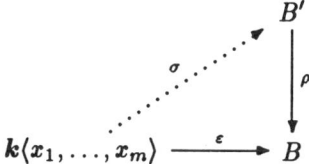

and $\rho : B'' \longrightarrow B$) and $\varphi = \rho_0 \times \varepsilon''$. This will prove the existence of $p_1, \ldots, p_m$ because the above equations for p_i are equivalent to the equation $\varphi(p_i) = (x_i, f(\varepsilon'(x_i)))$ in $\boldsymbol{k}\langle x_1, \ldots, x_m\rangle \times_B B''$. In order to establish the surjectivity of φ we first note that ρ_0 induces a surjection $\rho_0 : I'' \longrightarrow I$. Indeed we have a diagram

$$\begin{array}{ccccccccc} 0 & \longrightarrow & I'' & \longrightarrow & A\langle x_1, \ldots, x_m\rangle & \longrightarrow & B'' & \longrightarrow & 0 \\ & & \downarrow & & \downarrow & & \downarrow & & \\ 0 & \longrightarrow & I & \longrightarrow & \boldsymbol{k}\langle x_1, \ldots, x_m\rangle & \longrightarrow & B & \longrightarrow & 0 \end{array}$$

Reducing the top line modulo $\mathfrak{m}$ we obtain a diagram

$$\begin{array}{ccccccccc} & & \overline{I''} & \longrightarrow & \boldsymbol{k}\langle x_1, \ldots, x_m\rangle & \longrightarrow & \overline{B''} & \longrightarrow & 0 \\ & & \downarrow & & \downarrow & & \downarrow & & \\ 0 & \longrightarrow & I & \longrightarrow & \boldsymbol{k}\langle x_1, \ldots, x_m\rangle & \longrightarrow & B & \longrightarrow & 0 \end{array}$$

where the two right vertical arrows are isomorphisms. Hence $\overline{I''}$ surjects onto I and so I'' surjects onto I (if we assume the flatness of B'' then $\overline{I''} \longrightarrow I$ is an isomorphism).

We now prove that φ is into. Let $(q, b'') \in \boldsymbol{k}\langle x_1, \ldots, x_m\rangle \times_B B''$. Put $b = \varepsilon(q) = \rho(b'')$. Choose $p \in A\langle x_1, \ldots, x_m\rangle$ with $\varepsilon''(p) = b''$. Then $\varepsilon(\rho_0(p)) = \rho(\varepsilon''(p)) = \rho(b'') = b$. Hence $\varepsilon(q - \rho_0(p)) = 0$ and so $q - \rho_0(p) \in I$. Since $\rho_0 : I'' \longrightarrow I$ is onto we may choose $r'' \in I''$ such that $\rho_0(r'') = q - \rho_0(p)$. But then

$$\varphi(p + r'') = (\rho_0(p + r''), \varepsilon''(p + r'')) = (q, \varepsilon''(p)) = (q, b'').$$

□

Remark. Lemma 5.5 and 5.6 generalize to the case in which $\boldsymbol{k}\langle x_1, \ldots, x_m\rangle$ is replaced by the structure sheaf $\mathcal{O}_W$ of a Stein domain W in $\mathbb{C}^n$ and ε' and ε'' are surjections of $\mathcal{O}_W$–algebras. The generalization of Lemma 5.6 is carried out in detail in the first part of Lemma 6.3.

We now prove that h is surjective on isomorphism classes. We will in fact prove a more general *relative* version which we will need in the next chapter. Let S be a differential graded algebra over $\boldsymbol{k}$ and $\alpha : S \longrightarrow B$ be a surjective homomorphism of differential graded algebras. Let R be a resolvent for B over S. Thus R is a free S–algebra via the structure map $\iota : S \longrightarrow R$ and there is a commutative diagram

$$\begin{array}{ccc} S & \xrightarrow{\iota} & R \\ & {\scriptstyle\alpha}\searrow \quad \swarrow{\scriptstyle\varepsilon} & \\ & B & \end{array}$$

with ε a surjective quasi-isomorphism of differential graded algebras. We will assume that $S^0 = R^0$. In the case we are treating in this chapter $S = \boldsymbol{k}\langle x_1, \ldots, x_m\rangle$, α is the augmentation and ι is the inclusion.

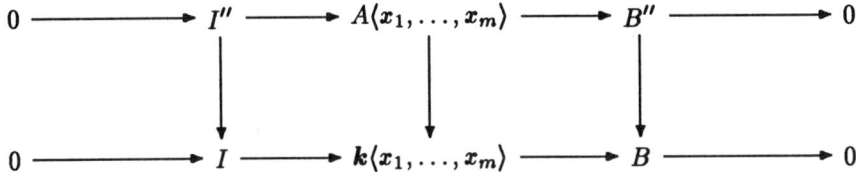

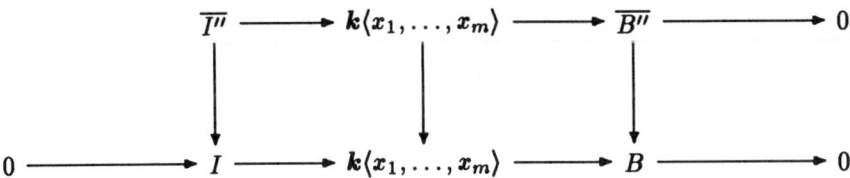

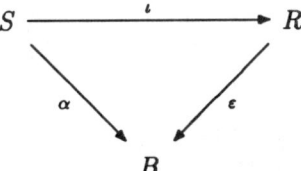

Since R is fixed we may assume R is constructed as in [F], Theorem 1.4. In this case there is a filtration of R by free differential graded S–algebras

$$S = R_0 \subset R_1 \subset \cdots \subset R$$

with compatible homomorphisms $\varepsilon_k : R_k \longrightarrow B$ such that ε_0 is surjective and ε_k, $k \geq 1$, induces an isomorphism on cohomology of degree less than k. Furthermore if $k > 0$ (resp. $k = 0$) then R_{k+1} is obtained from R_k as follows. Choose finitely many elements $x_\sigma \in Z^k(R_k)$ (resp. $\ker \varepsilon | S^0$) whose images generate $H^k(R_k)$ (resp. $\ker \varepsilon | S^0$) as an R_k^0–module. Then $R_{k+1} = R_k[e_\sigma]_\sigma$ with $\partial e_\sigma = x_\sigma$.

Now let $A \in \text{Obj } \mathcal{A}$ with maximal ideal $\mathfrak{m}$. We assume that we are given a flat deformation $\alpha' : S' \longrightarrow B'$ of the map $\alpha : S \longrightarrow B$ with base A. Thus S' and B' are flat deformations of S and B with base A. It follows (as in Lemma 5.6) that α' is surjective.

Lemma 5.8. *There exists* $R' \in \text{Def}(R; A)$ *and morphisms* $\iota' : S' \longrightarrow R'$, $\varepsilon' : R' \longrightarrow B'$ *such that* R' *is free over* S', $\iota' \equiv \iota \bmod \mathfrak{m}$, $\varepsilon' \equiv \varepsilon \bmod \mathfrak{m}$ *and* ε' *is a surjective quasi-isomorphism.*

Proof. We will construct inductively a sequence of free differential graded S'–algebras $S' = R'_0 \subseteq R'_1 \subseteq \cdots \subseteq R'$ flatly deforming the sequence $S = R_0 \subseteq R_1 \subseteq \cdots \subseteq R$, with compatible deformations $\varepsilon'_k : R'_k \longrightarrow B'$ of $\varepsilon_k : R_k \longrightarrow B$ such that $\varepsilon'_0 = \alpha'$ and if $k > 0$ then $\varepsilon'_k : R'_k \longrightarrow B'$ induces an isomorphism of cohomology of degree less than k. Moreover if $k \geq 1$ we will prove by induction that the group of $(k-1)$–cycles $Z^{k-1}(R'_k)$ is a flat A–module and that $\rho : R'_k \longrightarrow R_k$ induces an isomorphism $\overline{Z^{k-1}(R'_k)} \longrightarrow Z^{k-1}(R_k)$. Recall $^-$ denotes reduction module $\mathfrak{m}$.

We first claim that $\ker \varepsilon'_0 | (S')^0$ is a flat A–module and that $\rho : S' \longrightarrow S$ (reduction modulo $\mathfrak{m}$) induces an isomorphism $\overline{\ker \varepsilon'_0 | (S')^0} \longrightarrow \ker \varepsilon_0 | S^0$. The flatness part is immediate from the short exact sequence $\ker \varepsilon'_0 | (S')^0 \longrightarrow (S')^0 \longrightarrow B'$. We obtain a map of short exact sequences

$$\begin{array}{ccccccccc}
0 & \longrightarrow & \overline{\ker \varepsilon'_0 | (S')^0} & \longrightarrow & \overline{(S')^0} & \longrightarrow & \overline{B'} & \longrightarrow & 0 \\
 & & \downarrow & & \downarrow & & \downarrow & & \\
0 & \longrightarrow & \ker \varepsilon'_0 | (S')^0 & \longrightarrow & S^0 & \longrightarrow & B & \longrightarrow & 0
\end{array}$$

Since the two vertical arrows on the right are isomorphisms the claim follows. Now let $\{x_\sigma\} \subset \ker \varepsilon_0 | S^0$ be the elements described above which determine the passage from S to R_1. For each σ choose a lift of x_σ to $x'_\sigma \in \ker \varepsilon'_0 | (S')^0$. Define $R'_1 = S'[e_\sigma]$ with $\partial' e_\sigma = x'_\sigma$ and $\varepsilon'_1(e_\sigma) = 0$. It is easy to verify that R'_1 has the required properties. To continue the induction observe that $\ker \varepsilon'_0 | (s')^0$ flat implies $Z^1(R'_1)$ is flat and that ρ induces an isomorphism $\overline{Z^1(R'_1)} \longrightarrow Z^1(R_1)$.

Now assume that we have constructed R'_k as above. We now construct R'_{k+1}. We first claim that $Z^k(R'_k)$ is flat and the induced map $\rho : \overline{Z^k(R'_k)} \longrightarrow Z^k(R_k)$ is an isomorphism. Indeed let $C^k(R'_k)$ denote the A–module of elements of degree k and note that $H^{k-1}(R'_k) = 0$ by the induction hypothesis so $Z^{k-1}(R'_k) = B^{k-1}(R'_k)$

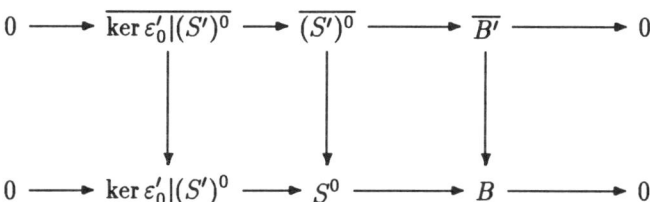

whence $B^{k-1}(R'_k)$ is flat. But we have a short exact sequence $0 \longrightarrow Z^k(R'_k) \longrightarrow C^k(R'_k) \longrightarrow B^{k-1}(R'_k) \longrightarrow 0$ so $Z^k(R'_k)$ is flat. We apply $\otimes_A \boldsymbol{k}$ to the previous short exact sequence and obtain a short exact sequence (since $B^{k-1}(R'_k)$ is flat)

$$0 \longrightarrow \overline{Z^k(R'_k)} \longrightarrow \overline{C^k(R'_k)} \longrightarrow \overline{B^{k-1}(R'_k)} \longrightarrow 0.$$

It follows from the 5-lemma that $\rho : \overline{Z^k(R'_k)} \longrightarrow Z^k(R_k)$ is an isomorphism.

We next consider the exact sequence $B^k(R_k) \longrightarrow Z^k(R_k) \longrightarrow H^k(R_k)$. Let $\{x_\sigma\}_\sigma$ be the elements in $Z^k(R_k)$ defined above such that $R_{k+1} = R_k[e_\sigma]_\sigma$ with $\partial e_\sigma = x_\sigma$. Choose lifts x'_σ of x_σ to $Z^k(R'_k)$. Define $R'_{k+1} = R'_k[e_\sigma]$ with $\partial e_\sigma + x'_\sigma$ and $\varepsilon'_{k+1}(e_\sigma) = 0$.

We now prove that R'_{k+1} has the correct properties. We note that $C^k(R'_{k+1}) = C^k(R'_k)$ whence $Z^k(R'_{k+1}) = Z^k(R'_k)$ so $Z^k(R'_{k+1})$ is flat over A and the induced map $\rho : \overline{Z^k(R'_{k+1})} \longrightarrow Z^k(R_{k+1})$ is an isomorphism (note $Z^k(R_{k+1}) = Z^k(R_k)$). Finally we show that $H^k(R'_{k+1}) = 0$. We apply $\otimes_A \boldsymbol{k}$ to the short exact sequence $0 \longrightarrow B^k(R'_{k+1}) \longrightarrow Z^k(R'_{k+1}) \longrightarrow H^k(R'_{k+1}) \longrightarrow 0$ to obtain a right exact sequence and a diagram

$$\begin{array}{ccccccccc}
 & & \overline{B^k(R'_{k+1})} & \longrightarrow & \overline{Z^k(R'_{k+1})} & \longrightarrow & \overline{H^k(R'_{k+1})} & \longrightarrow & 0 \\
 & & \downarrow & & \downarrow & & \downarrow & & \\
0 & \longrightarrow & B^k(R_{k+1}) & \longrightarrow & Z^k(R_{k+1}) & \longrightarrow & H^k(R_{k+1}) & \longrightarrow & 0
\end{array}$$

where the vertical arrows are induced by ρ. The middle vertical arrow is an isomorphism. Also $\rho : R'_{k+1} \longrightarrow R_{k+1}$ is onto by construction so the left vertical arrow is a surjection. Hence the right vertical arrow is an isomorphism. But $H^k(R_{k+1}) = 0$ and consequently $\overline{H^k(R'_{k+1})} = 0$ whence $H^k(R'_{k+1}) = 0$ by Nakayama's Lemma. □

Remark. In fact we will need a still more general version of Lemma 5.6 in Chapter 6. Let $(W, \mathcal{O}_W)$ be a Stein space. Then the conclusion of Lemma 5.6 remains true when S, R and B are $\mathcal{O}_W$–algebras with S a free graded-finite $\mathcal{O}_W$–algebra (i.e. generated in each degree by a finite number of global sections), S' and B' are $\mathcal{O}_W \otimes A$–algebras and R admits a filtration

$$S = R_0 \subset R_1 \subset \cdots \subset R$$

with R_{k+1} obtained from R_k as above with x_σ and e_σ global sections. The point is that the resolvent for a model space $X \subset \mathbb{C}^n$ has this property, [P], Theorem 2.1. To prove this new version prove inductively that $Z^{k-1}(R'_k)$ is a flat A–module that is a *coherent* $\mathcal{O}_W$–module and use Cartan's Theorem B.

We now wish to prove that the functor h is full. Again we prove a more general relative version that we will need in the next chapter. Let $B'' \in \text{Obj Def}(B; A)$

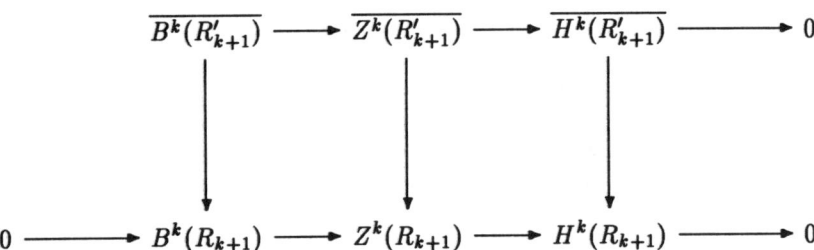

and $f : B' \longrightarrow B''$ be a morphism in $\mathrm{Def}(B; A)$. Let S be a differential graded algebra that is a free $\boldsymbol{k}$–algebra and $\alpha : S \longrightarrow B$ be a surjective homomorphism. Let $\varepsilon : R \longrightarrow B$ be a resolvent of B as an S–algebra. Hence there is a commutative diagram of morphisms of differential graded algebras

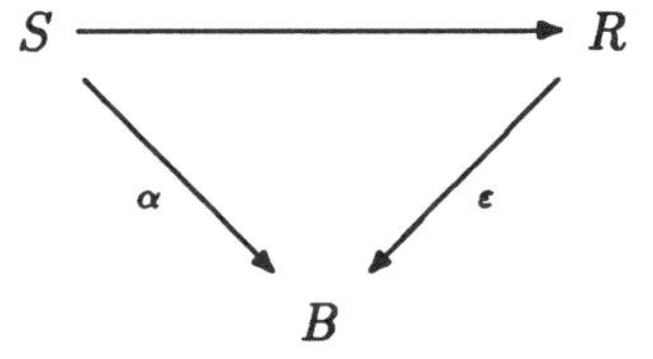

with ε a surjective quasi-isomorphism. Let $S', S'' \in \mathrm{Obj}\ \mathrm{Def}(S; A)$, this groupoid is defined in the obvious way, and $R', R'' \in \mathrm{Obj}\ \mathrm{Def}(R; A)$ such that R' is a free S'–algebra, R'' is a free S''–algebra and such that there are commutative triangles of morphisms of differential graded algebras

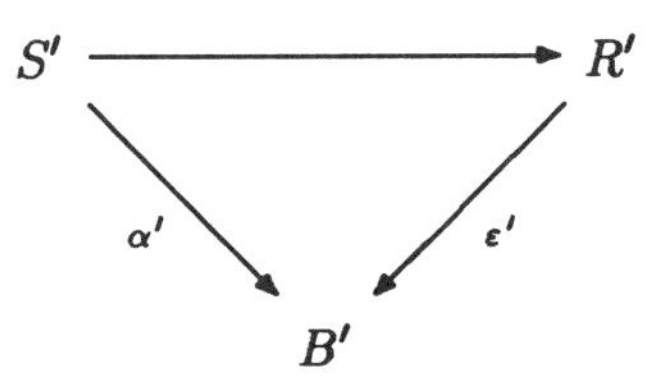

and

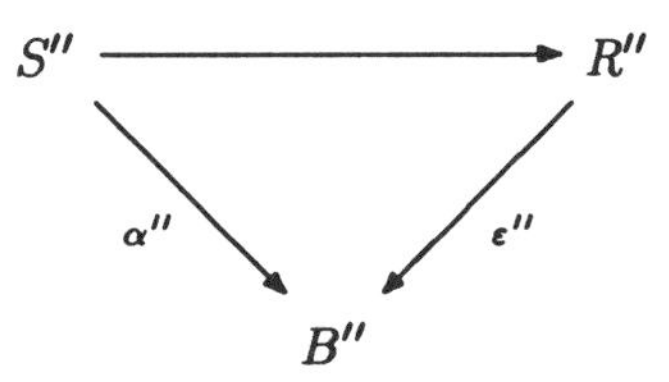

deforming the triangle above. Assume further that ε' and ε'' are surjective quasi-isomorphisms and that $(S')^0 = (R')^0$ and $(S'')^0 = (R'')^0$.

Lemma 5.9. *Suppose $g : S' \longrightarrow S''$ is a morphism in $\mathrm{Def}(S; A)$ over $f : B' \longrightarrow B''$. Then there exists $\tilde{g}$ a morphism in $\mathrm{Def}(R; A)$ such that $\tilde{g}|S' = g$.*

Proof. Let e be a generator of R' over S' of minimal degree. We may suppose $\deg e > 0$. Then $\partial' e \in s'$ and $g(\partial' e) = \partial'' p$ for some $p \in R''$. Since $g \equiv id \bmod \mathfrak{m}$, $\bar{p} - \bar{e}$ is a cycle in R and there exists $q \in R''$ such that $\bar{p} - \bar{e} = \partial \bar{q}$. Now define $\tilde{g} : S'[e] \longrightarrow R''$ by requiring that

$$\tilde{g}(e) = p - \partial'' q.$$

The lemma follows by induction. □

Remark. Again we leave the generalization to sheaves of free algebras finitely generated in each degree over the structure sheaf of a Stein space to the reader.

Proposition 5.1. *The natural transformation h induces an isomorphism of functors*

$$h : \mathrm{Iso}\ \mathrm{Def}(R; \cdot) \longrightarrow \mathrm{Iso}\ \mathrm{Def}(B; \cdot).$$

Proof. Let $A \in \mathrm{Obj}\ \mathcal{A}$ be given. Then the mapping

$$h : \mathrm{Iso}\ \mathrm{Def}(R; A) \longrightarrow \mathrm{Iso}\ \mathrm{Def}(B; A)$$

is surjective by Lemma 5.6. We claim h is injective. Indeed, suppose there exist $x_1, x_2 \in \mathrm{Def}(R; A)$ and an isomorphism $\varphi : h(x_1) \longrightarrow h(x_2)$. By Lemma 5.7, h is full and consequently there exists a morphism $\psi : x_1 \longrightarrow x_2$. But $\mathrm{Def}(R; A)$ is a groupoid so ψ is an isomorphism. □

Corollary. *The complete local* $\mathbf{k}$*–algebra* R_L *is a pro-representable hull for the functor* Iso Def$(B;\cdot)$. □

We have now achieved the goal of this section.

THEOREM A. *The local tangent complex* L *controls the deformation theory of the germ* $(V,0)$.

Proof. Let (X,x) be the parameter space for the versal deformation of $(V,0)$. The ring $\hat{\mathcal{O}}_{X,x}$ is a pro-representable hull for the functor Iso Def$(B;\cdot)$. But R_L is also a pro-representable hull for Iso Def$(B;\cdot)$. By the uniqueness of pro-representable hulls $R_L \cong \hat{\mathcal{O}}_{X,x}$. □

We now justify why we have named the differential graded Lie algebra L above the *local* tangent complex instead of the infinitesimal tangent complex. Let V be an analytic subset of $\mathbb{C}^N$ with $0 \in V$ and X denote the intersection of V with a small closed ball W centered at the origin. Let $i : X \longrightarrow W$ be the inclusion and $\mathcal{O}_W$ (resp. $\mathcal{O}_X$) denote the sheaf $\mathcal{O}_{\mathbb{C}^N}|W$ (resp. $\mathcal{O}_V|X$). Using Cartan's Theorems A and B one can easily construct a resolvent R for the sheaf $i_*\mathcal{O}_X$ as an $\mathcal{O}_W$–module, see the beginning of the proof of Theorem 3.1 in [P]. We define $L_X = \mathrm{Der}^+(R,R)$. Thus we are considering the special case of the situation of Chapter 4 in which the covering $\mathcal{K}$ consists of the single set X. We let $\mathcal{L}^\bullet = \mathcal{D}er^+(R,R)$ denote the corresponding sheaf of derivations and $\mathcal{T}^\bullet$ denote the cohomology sheaves of the complex $\mathcal{L}^\bullet$.

Now choose $x \in X$. We have a natural map of stalks

$$\rho_x : \mathcal{D}er^+(R,R)_x \longrightarrow \mathrm{Der}^+(R_x,R_x).$$

Lemma 5.10. *The map* ρ_x *is a quasi-isomorphism of differential graded Lie algebras.*

Proof. The augmentation $\varepsilon : R \longrightarrow i_*\mathcal{O}_X$ gives rise to a diagram

$$\begin{array}{ccc} \mathcal{D}er^+(R,R)_x & \longrightarrow & \mathcal{D}er^+(R,i_*\mathcal{O}_X)_x \\ \downarrow & & \downarrow \\ \mathrm{Der}^+(R_x,R_x) & \longrightarrow & \mathrm{Der}^+(R_x,\mathcal{O}_{X,x}) \end{array}$$

By applying Lemma 4.3 to a cofinal system of Stein neighborhoods of x in X one finds that the upper horizontal arrow is a quasi-isomorphism. By [P], Proposition 1.10 (the last three paragraphs of the proof) the lower horizontal arrow is a quasi-isomorphism. Hence it suffices to prove that the right vertical arrow is a quasi-isomorphism. By Lemma 4.4 it suffices to prove that the stalk map

$$\rho_x : \mathrm{Hom}_{\mathcal{O}_X}(\mathcal{P},\mathcal{O}_X)_x \longrightarrow \mathrm{Hom}_{\mathcal{O}_{X,x}}(\mathcal{P}_x,\mathcal{O}_{X,x})$$

is a quasi-isomorphism. But this latter map is clearly an isomorphism. □

Corollary. *T^i is supported on the singular set of X for $i > 0$.* □

We next observe that we have a natural map (taking the stalk of global sections) $\sigma_x : L_X^{\cdot} \longrightarrow \mathcal{D}er^+(R,R)_x$ whence a map $r_x : L_X^{\cdot} \longrightarrow \mathrm{Der}^+(R_x, R_x)$. Now suppose V has a unique singular point which is located at 0. We have seen that $L_{V,0}^{\cdot} = \mathrm{Der}^+(R_0, R_0)$ is a controlling differential graded Lie algebra for deformations of $\mathcal{O}_{V,0}$ hence for deformations of the germ $(V,0)$. For the definition of the fibered groupoid $\mathrm{Def}(X;\cdot)$ see Chapter 6.

Theorem 5.1. *The map $r_0 : L_X \longrightarrow L_{V,0}$ is a one-quasi-isomorphism.*

Proof. Since X is Stein, we have $H^i(L_X) = \Gamma(X, \mathcal{T}^i)$. Thus $H^i(r_0)$ is the germ map $\Gamma(X, \mathcal{T}^i) \longrightarrow \mathcal{T}_0^i$. This is an isomorphism for $i > 0$ since $\mathcal{T}^i$ is concentrated at 0 for $i > 0$. □

Corollary. *$L_{V,0}$ controls the deformation theory of X.*

Proof. We will see (as a special case of Theorem C) that L_X controls the deformation theory of X. □

We can now prove Theorem B of the introduction.

THEOREM B. *L_V is one-quasi-isomorphic to $L_{V,0}$.*

Proof. It suffices to prove that L_V and L_X are one-quasi-isomorphic. Choose a cover $\{B_i\}_{i \in I}$ of $\mathbb{C}^N$ by open balls. Assume B_{i_0} is centered at 0 for some $i_0 \in I$, that $X = \bar{B}_{i_0} \cap V$ and that $0 \notin \bar{B}_i$, $i \neq i_0$. We let $V_i = B_i \cap V$ and $\mathcal{V} = \{V_i\}$. We assume that $\mathrm{Nerve}\{\bar{V}_i\} = \mathrm{Nerve}\,\mathcal{V}$ and put $\mathcal{N} = \mathrm{Nerve}(\mathcal{V})$. We obtain simplicial schemes of Stein compacta X_*, W_* as in Chapter 4 with $X_i = \bar{V}_i$ whence $X_{i_0} = X$ and $W_i = \bar{B}_i$ whence $W_{i_0} = \bar{B}_{i_0}$. We have a (restriction) map of double complexes

$$C^{\cdot}\left(\mathrm{sd}\,\mathcal{N}, \mathrm{Hom}(\Omega^1_{R_V}, R_V)\right) \longrightarrow C^{\cdot}\left(\mathrm{sd}\{i_0\}, \mathrm{Hom}(\Omega^1_{R_x}, R_x)\right).$$

By Theorem 4.3 it suffices to prove that the restriction map induces isomorphisms on cohomology groups of the total complex of positive degree. By Theorem 4.4 the E_2–term of the second spectral sequence (taking tangent cohomology first) for the first double complex is concentrated in (simplicial) degree zero since V and X are Stein and $\mathcal{T}^{\cdot}$ is coherent. But the map of $E_2^{0,q}$–terms is just the restriction map $\Gamma(V, \mathcal{T}^q) \longrightarrow \Gamma(X, \mathcal{T}^q)$. □

6. The Global Tangent Complex Controls the Flat Deformations of a Complex Analytic Space

Let X be a complex analytic space and L_X be the tangent complex of X (see Chapter 4). The purpose of this chapter is to prove the following theorem.

THEOREM C. *The differential graded Lie algebra L_X controls the deformation theory of X, more precisely, if there exists a formally versal deformation of X then its base is isomorphic to R_{L_X}.*

Corollary. *Let X be a Stein analytic space with a unique singular point $x \in X$. Put $U = X - \{x\}$ and assume* $\operatorname{depth}_{\{x\}}\mathcal{O}_X \geq 3$. *Let R be a resolvent for $\mathcal{O}_{X,x}$ and $L_{X,x} = \operatorname{Der}^+(R, R)$ be the local tangent complex of the germ (X, x). Let L_U be the tangent complex of U. Then the complete local $\mathbb{C}$–algebras $R_{L_{X,x}}$ and R_{L_U} are isomorphic.*

Proof. In [Sc2] it is proved that if $\operatorname{depth}_{\{x\}}X \geq 3$ then U has a formal versal deformation (this is equivalent to $T^1(U) = \dim H^1(L_U) < \infty$ by [Be]) and the formal deformation theories of (X, x) and U are isomorphic. □

Remark. See [Ar2], §9, for a detailed discussion of Schlessinger's Theorem.

We begin by describing the fibered groupoid $\operatorname{Def}(X; \cdot)$ of flat deformations with infinitesimal base of the complex analytic space X. Let $A \in \operatorname{Obj} \mathcal{A}$ with maximal ideal $\mathfrak{m}$. An object $(X'; \pi')$ of $\operatorname{Def}(X; A)$ is a complex analytic space X' and a flat map $\pi' : X' \longrightarrow \operatorname{Spec} A$ such that the following diagram is cartesian

$$\begin{array}{ccc} X & \longrightarrow & X' \\ {\scriptstyle\pi}\downarrow & & \downarrow{\scriptstyle\pi'} \\ \operatorname{Spec}\mathbb{C} & \longrightarrow & \operatorname{Spec} A \end{array}$$

A morphism $f : (X', \pi') \longrightarrow (X'', \pi'')$ is a morphism of analytic spaces $f : X' \longrightarrow X''$ such that $f|X = id$ and the diagram

$$\begin{array}{ccc} X' & \xrightarrow{\ f\ } & X'' \\ & {\scriptstyle\pi'}\searrow \quad \swarrow{\scriptstyle\pi''} & \\ & \operatorname{Spec} A & \end{array}$$

commutes. We observe that (since the underlying space of X' is X) an object (X', π') of $\operatorname{Def}(X; A)$ gives rise to and is determined by a sheaf B' of flat A–algebras on X and a cocartesian square

$$\begin{array}{ccc} B' & \longrightarrow & \mathcal{O}_X \\ \downarrow & & \downarrow \\ A & \longrightarrow & \mathbb{C} \end{array}$$

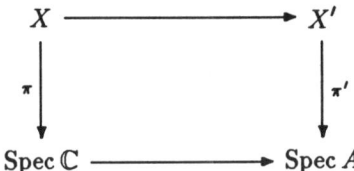

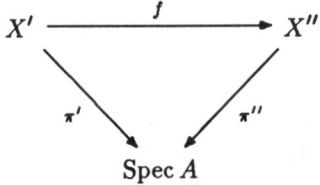

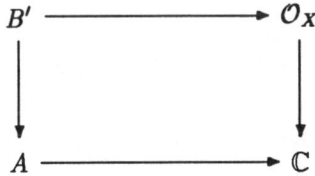

We note that B' is necessarily a sheaf of analytic A–algebras. The correspondence between B' and X' is given by the formula $\mathcal{O}_{X'} = B'$.

We will need the following gluing lemma for simplicial sheaves. Let $\mathcal{U} = \{U_i\}_{i\in I}$ be a cover of X and U_* be the corresponding simplicial topological space on Nerve($\mathcal{U}$). We let k^* be the functor from the category of sheaves on X to the category of simplicial sheaves on U_* given by $(k^*F)_\alpha = F|U_\alpha$ with structure maps $\mu_{\beta\alpha} : (k^*F_\alpha)|U_\beta = F|U_\beta \longrightarrow (k^*F)_\beta = F|U_\beta$ for $\alpha \subset \beta$ given by the corresponding identity maps.

Lemma 6.1. *Let F_* be a simplicial sheaf of abelian groups on U_* such that all the structure maps $\mu_{\beta\alpha} : F_\alpha|U_\beta \longrightarrow F_\beta$, $\alpha \subset \beta$, are isomorphisms. Then there exists a sheaf F on X and isomorphisms $\eta_\alpha : F|U_\alpha \longrightarrow F_\alpha$ such that $\mu_{\beta\alpha} = \eta_\beta \circ (\eta_\alpha|U_\beta)^{-1}$. Moreover the sheaf F is unique up to isomorphism and the η_α's are determined up to right multiplication by a unique $\tau \in \operatorname{Aut} F$.*

Proof. Let i, j be vertices of $\mathcal{N}$. Define $\theta_{ji} : F_i|U_i \cap U_j \longrightarrow F_j|U_i \cap U_j$ by $\theta_{ji} = \mu_{ij,j}^{-1} \circ \mu_{ij,i}$ (note that $\mu_{ij,j} : F_j|U_i \cap U_j \longrightarrow F_{ij}$ is an isomorphism by hypotheses). From the identity $\mu_{\gamma\beta} \circ \mu_{\beta\alpha} = \mu_{\gamma\alpha}$ for $\alpha \subset \beta \subset \gamma$ it follows immediately that $\theta_{ij}\,\theta_{jk} = \theta_{ik}$, for $i, j, k \in I$ with $U_i \cap U_j \cap U_k \neq \emptyset$. We define a sheaf F on X by defining $F(U)$ to be the collection $\{s_k\}_{k\in I}$ with $s_k \in F(U \cap U_k)$ satisfying the compatibility conditions $s_k = \theta_{kj}(s_j)$ on $U \cap U_j \cap U_k$. We define $\eta_i(U)$, $i \in I$, by $\eta_i(\{s_k\}) = s_i$. Finally given $\beta \in \mathcal{N}$ choose a vertex i of β and define $\eta_\beta = \mu_{\beta i} \circ \eta_i|U_\beta$. To check that η_β does not depend on i let j be another vertex of β. Then

$$\begin{aligned}\mu_{\beta j} \circ \eta_j &= \mu_{\beta,ij}\mu_{ij,j}\eta_j = \mu_{\beta,ij}\mu_{ij,j}\theta_{ji}\eta_i \\ &= \mu_{\beta,ij}\mu_{ij,i}\eta_i = \mu_{\beta i} \circ \eta_i.\end{aligned}$$

We can now prove the formula $\mu_{\beta\alpha} = \eta_\beta \circ \eta_\alpha^{-1}$ for $\alpha \subset \beta$. Indeed let i be a vertex of α. We have $\eta_\alpha = \mu_{\alpha i}\eta_i$ on U_α (and hence on U_β) and $\eta_\beta = \mu_{\beta i}\eta_i$. From the first formula we obtain $\mu_{\beta\alpha}\eta_\alpha = \mu_{\beta\alpha}\mu_{\alpha i}\eta_i = \mu_{\beta i}\eta_i = \eta_\beta$.

To prove uniqueness we note that the isomorphisms $\{\eta_\alpha\}$ fit together to give an isomorphism $\eta_* : k^*F \longrightarrow F_*$ of simplicial sheaves on U_*. Thus if $\{\eta'_\alpha\}$ is another system satisfying the above equations then we obtain $\eta'_* : k^*F \longrightarrow F_*$ and $\tau_* \in \operatorname{Aut}(k^*F)$ defined by $\tau_* = (\eta'_*)^{-1} \circ \eta_*$. But clearly the functor k^* is full whence $\tau_* = k^*(\tau)$ with $\tau \in \operatorname{Aut}(F)$. □

We now return to the set-up of Chapter 4. Thus we have simplicial complex spaces $U_* \subset X_* \subset V_*$ and W_* with W_* smooth and a proper embedding $\tilde{i} : V_* \longrightarrow W_*$ restricting to $i : X_* \longrightarrow W_*$. We also have a resolvent R for $i_*\mathcal{O}_{X_*}$ that we fix once and for all. Now let X' be a deformation of X and B' be the corresponding sheaf of analytic A–algebras on X. We obtain a simplicial sheaf B'_* on X_* by the formula $B'_\alpha = B'|X_\alpha$, $\alpha \in N$. The structure maps $\mu_{\beta\alpha} : j^*_{\alpha\beta}B'_\alpha \longrightarrow B'_\beta$, $a \subset \beta$, are the identity maps. We may regard B'_* as a deformation of X_* — we leave the formal definitions in this case to the reader.

We now prove the global analogues of Lemmas 5.6 and 5.7. We fix W_* and a surjective homomorphism $\varepsilon : \mathcal{O}_{W_*} \longrightarrow i_*\mathcal{O}_{X_*}$. Let $A \in \operatorname{Obj} \mathcal{A}$ with maximal ideal $\mathfrak{m}$. We define the groupoid $\operatorname{Emb}\operatorname{Def}(X; A)$ of embedded deformations of X as follows. An object of $\operatorname{Emb}\operatorname{Def}(X; A)$ is a deformation $\rho : B' \longrightarrow \mathcal{O}_X$ of X and a surjective homomorphism $\varepsilon' : \mathcal{O}_{W_*} \otimes A \longrightarrow i_* B'_*$ such that the following diagram commutes

$$\begin{array}{ccc} \mathcal{O}_{W_*} \otimes A & \xrightarrow{\rho_0} & \mathcal{O}_{W_*} \\ \big\downarrow & & \big\downarrow \\ i_* B'_* & \xrightarrow{\rho} & i_* \mathcal{O}_{X_*} \end{array}$$

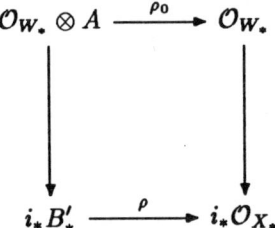

Here ρ_0 is the natural map induced by the reduction map $A \longrightarrow \mathbb{C}$. Now let $\varepsilon' : \mathcal{O}_{W_*} \otimes A \longrightarrow i_* B'_*$ and $\varepsilon'' : \mathcal{O}_{W_*} \otimes A \longrightarrow I_* B''_*$ be objects in $\operatorname{Emb}\operatorname{Def}(X; A)$. Then a morphism from ε' to ε'' is a pair (φ, ψ) where $\varphi : \mathcal{O}_{W_*} \otimes A \longrightarrow \mathcal{O}_{W_*} \otimes A$ and $\psi : i_* B'_* \longrightarrow i_* B''_*$ are morphisms of deformations (i.e. morphisms of simplicial sheaves of A–algebras reducing to the identity modulo $\mathfrak{m}$) such that the following diagram commutes

$$\begin{array}{ccc} \mathcal{O}_{W_*} \otimes A & \xrightarrow{\varphi} & \mathcal{O}_{W_*} \otimes A \\ \big\downarrow{\scriptstyle \varepsilon'} & & \big\downarrow{\scriptstyle \varepsilon''} \\ i_* B'_* & \xrightarrow{\psi} & i_* B''_* \end{array}$$

Clearly we have a functor Φ from $\operatorname{Emb}\operatorname{Def}(X; A) \longrightarrow \operatorname{Def}(X; A)$ given by $\Phi(\varepsilon' : \mathcal{O}_{W_*} \otimes A \longrightarrow i_* B'_*) = B'$ and $\Phi(\varphi, \psi) = \psi$. We now prove that Φ is surjective on isomorphism classes and full (clearly it is not faithful in general). The next proof follows that of [F], Lemma 3.13.

Lemma 6.2. *The functor Φ is surjective on isomorphism classes.*

Proof. We first push forward the deformation B'_* of X_* to one of W_*. This we do by defining a simplicial sheaf C_* of analytic A–algebras on W_* by $C_* = \mathcal{O}_{W_*} \times_{i_* \mathcal{O}_{X_*}} i_* B'_*$. We define the structure maps $p_{\alpha\beta}^{-1} C_\beta \longrightarrow C_\alpha$ as the fiber product of those of $\mathcal{O}_{W_*}$ and $i_* B'_*$. We obtain a commutative diagram of simplicial sheaves of algebras

$$\begin{array}{ccccccccc} 0 & \longrightarrow & i_* I_* & \longrightarrow & C_* & \longrightarrow & \mathcal{O}_{W_*} & \longrightarrow & 0 \\ & & \big\downarrow & & \big\downarrow & & \big\downarrow & & \\ 0 & \longrightarrow & i_* I_* & \longrightarrow & i_* B'_* & \longrightarrow & i_* \mathcal{O}_{X_*} & \longrightarrow & 0 \end{array}$$

It follows easily from [Schu], Theorem 2.6, that W_*, equipped with the structure sheaf C_*, is again a simplicial scheme of Stein compacts which we denote W'_*. The above diagram of sheaves is dual to a closed embedding $\iota : W_* \longrightarrow W'_*$ defined by the ideal sheaf $i_* I_*$.

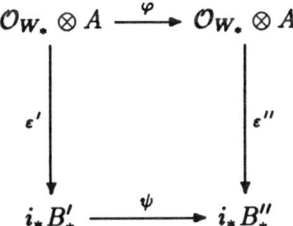

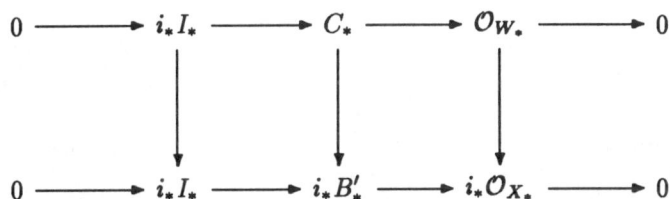

We now construct a cross-section $\sigma : \mathcal{O}_{W_*} \longrightarrow \mathcal{O}_{W'_*} = C_*$. Such a cross-section is equivalent to a morphism $p : W'_* \longrightarrow W_*$ such that $p \circ \iota = id$. We first assume that the ideal sheaf i_*I_* has square zero. In this case the desired cross-section is constructed in [F], Lemma 3.13, using the projectivity of $\Omega^1_{W_*}$ in $\mathrm{Coh}(\mathcal{O}_{W_*})$. In fact, Flenner proves that any extension $\iota : W_* \longrightarrow W'_*$ defined by an ideal sheaf N_* of square zero admits a left inverse. Now put $J = i_*I_*$ and consider the sequence $\{W_k\}$ of simplicial schemes of Stein compacta such that $W_k = (W_*, C_*/J^k)$. There exists n such that $\mathfrak{m}^n = 0$ whence $J^n = 0$ and $W_n = W'_*$. We obtain a sequence of closed embeddings

$$W_* = W_1 \longrightarrow W_2 \longrightarrow \cdots \longrightarrow W_{n-1} \longrightarrow W_n = W'_*$$

with $W_{k-1} \subset W_k$ defined by an ideal of square zero. We construct a left inverse $p_k : W_k \longrightarrow W_1$ inductively on k as in [Schu], Theorem 4.3, using the above result of Flenner.

We obtain an induced map of A-algebras $\mathcal{O}_{W_*} \otimes A \longrightarrow C_*$ whence a map $\varepsilon' : \mathcal{O}_{W_*} \otimes A \longrightarrow i_*B'_*$. Then $\varepsilon' \equiv \varepsilon \bmod \mathfrak{m}$ whence ε' is surjective by Nakayama's Lemma. □

We will adopt the following notation throughout the next lemma. If $f : (X, \mathcal{O}_X) \longrightarrow (Y, \mathcal{O}_Y)$ is a morphism of ringed spaces then $|f|$ will denote the underlying map of topological spaces and $f^* : \mathcal{O}_Y \longrightarrow |f|_*\mathcal{O}_X$ will denote the structure morphism. In case $X = Y$ and $|f| = id$ then $f : (X, \mathcal{O}_X) \longrightarrow (X, \mathcal{O}_Y)$ is determined by the morphism f^* of sheaves of algebras over X.

Remark. In the proof of the next lemma we will need two observations concerning maps of analytic subspaces of open subsets of $\mathbb{C}^m$. First if X' and X'' are two such subspaces and $f : X' \longrightarrow X''$ is a morphism then f is determined by the m global sections $a_i = f^*z''_i$, $1 \le i \le m$, in $\Gamma(X', \mathcal{O}_{X'})$ where z''_i is the restriction of the coordinate function z_i of $\mathbb{C}^m$ to X'', $1 \le i \le m$, (see [Fi], 0.19). Second suppose A is an Artin local $\mathbb{C}$-algebra, W is an open subset of $\mathbb{C}^m$ and $\widetilde{W} = W \times \operatorname{Spec} A$. Let $a_1, a_2, \ldots, a_m$ be global sections of $\mathcal{O}_{\widetilde{W}}$ such that $a_i|W = z_i$, $1 \le i \le m$. Then there exists a unique morphism $F : \widetilde{W} \longrightarrow \widetilde{W}$ such that $|F| = id$ and $F^*z_i = a_i$, $1 \le i \le m$, (see [Fi], 0.17 and 0.19). Here we have written z_i instead of p^*z_i where $p : \widetilde{W} \longrightarrow W$ is the projection.

Lemma 6.3. *The functor Φ is full.*

Proof. Let $\varepsilon' : \mathcal{O}_{W_*} \otimes A \longrightarrow i_*B'_*$ and $\varepsilon'' : \mathcal{O}_{W_*} \otimes A \longrightarrow i_*B''_*$ be as above and $f^* : i_*B'_* \longrightarrow i_*B''_*$ be a morphism of simplicial sheaves of A-algebras. We are required to lift f^* to a morphism $F^* : \mathcal{O}_{W_*} \otimes A \longrightarrow \mathcal{O}_{W_*} \otimes A$ of simplicial sheaves of A-algebras such that F reduces to the identity modulo $\mathfrak{m}$. We first solve the local case (i.e. the case in which the underlying complex $\mathcal{N}$ consists of a single point). It is more convenient to consider spaces rather than algebras in this case.

Let W be a Stein domain in $\mathbb{C}^m$ and $X \subset W$ be the analytic subset defined by an ideal sheaf $I \subset \mathcal{O}_W$ with $i : X \longrightarrow W$ the inclusion and let $j : W \longrightarrow \widetilde{W}$ be the inclusion. Thus we are given deformations X' and X'' of X, a morphism $f : X' \longrightarrow X''$ (so $f|X = id$) and embeddings $i' : X' \longrightarrow \widetilde{W}$, $i'' : X'' \longrightarrow \widetilde{W}$ determined by ideal sheaves I' and I''. We let $j' : X \longrightarrow X'$ and $j'' : X \longrightarrow X''$ be the inclusions whence $(j')^* : I' \longrightarrow I$ and $(j'')^* : I' \longrightarrow I$ are surjective homomorphisms of sheaves over X (it suffices to prove this on stalks and the required proof is contained in the proof of Lemma 5.7). We are required to extend f to $F : \widetilde{W} \longrightarrow \widetilde{W}$ such that $F|W = id$.

Let $z_1, \ldots, z_m$ be the coordinate functions on $\mathbb{C}^m$ and $z_1', \ldots, z_m'$ (resp. $z_1'', \ldots, z_m''$) be the global sections of $\mathcal{O}_{X'}$ (resp. $\mathcal{O}_{X''}$) determined by restriction. We note that the underlying space map $|f|$ of f is the identity and $f^* : \mathcal{O}_{X''} \longrightarrow |f|_*\mathcal{O}_{X'} = \mathcal{O}_{X'}$ is a homomorphism of sheaves of A-algebras over X. Put $a_i = f^*z_i''$; $1 \leq i \leq m$, and observe that the restrictions of a_i and $z_i \in \Gamma(W, \mathcal{O}_W)$ to X coincide, $1 \leq i \leq m$, since $f|X = id$. We now recall the definition of the "sum" analytic space $Y' = W \amalg_X X'$ as given in [Schu], Theorem 2.7, namely $(Y', \mathcal{O}_Y) = (W, \mathcal{O}_W \times_{|i|_*\mathcal{O}_X} |i|_*\mathcal{O}_{X'})$. Since Y' is the categorical sum there is a natural map $k : Y' \longrightarrow \widetilde{W}$ given by $k = j \amalg i$. We claim that k is a closed embedding. On the space level there is no problem since $|k| = j$. It remains to prove that $k^* : \mathcal{O}_{\widetilde{W}} \longrightarrow \mathcal{O}_{Y'}$ is onto. But this is equivalent to proving that all the stalk maps associated to k^* are onto. Let $x \in X$. We may identify $\mathcal{O}_{\widetilde{W}_*,x}$ with $A\langle z_1, \ldots, z_m\rangle$ and $\mathcal{O}_{W_*,x}$ with $\mathbb{C}\langle z_1, \ldots, z_m\rangle$. Then k_x^* is identified with the map $\varphi : A\langle z_1, \ldots, z_m\rangle \longrightarrow \mathbb{C}\langle z_1, \ldots, z_m\rangle \times_B B'$ of the proof of Lemma 5.7 where $B = \mathcal{O}_{X,x}$ and $B' = \mathcal{O}_{X',x}$. Since we proved φ was onto in Lemma 5.7 we find that k^* is onto.

We can now construct F. Indeed the pairs (a_i, z_i), $1 \leq i \leq m$, determine global sections b_i of $\mathcal{O}_{Y'}$. By Cartan's Theorem B we may extend b_i to $\tilde{b}_i \in \Gamma(\widetilde{W}, \mathcal{O}_{\widetilde{W}})$, $1 \leq i \leq m$. Now a morphism $K : \widetilde{W} \longrightarrow W$ over $\operatorname{Spec}\mathbb{C}$ satisfying $K^*z_i = \tilde{b}_i$, $1 \leq i \leq m$, gives rise to a unique morphism $F : \widetilde{W} \longrightarrow \widetilde{W}$ satisfying $F^*z_i = \tilde{b}_i$ whence $F|X' = f$ and $F|W = id$. Thus it suffices to prove that K as above exists. But this follows from the above remark. For comparison with the following global argument note that this paragraph may be viewed as solving the extension problem

$$\begin{array}{ccc} \widetilde{W} & \overset{F}{\cdots\cdots\rightarrow} & \widetilde{W} \\ \uparrow & & \uparrow \\ Y' & \overset{H}{\longrightarrow} & Y'' \end{array}$$

where $H : Y' = W \amalg_X X' \longrightarrow Y'' = W \amalg_X X''$ is given by $H = id \amalg f$ whence $H^*c_i = b_i$, where $c_i = (z_i, z_i'') = z_i|Y''$, $1 \leq i \leq m$.

We now treat the case of general $\mathcal{N}$. By the previous analysis for every $\alpha \in \mathcal{N}$ we have $G_\alpha : \widetilde{W}_\alpha \longrightarrow \widetilde{W}_\alpha$ with $G_\alpha|W_\alpha = id$ and $G_\alpha|X_\alpha' = f_\alpha$. We need to modify $\{G_\alpha\}$ to $\{F_\alpha\}$ so that for every $\alpha, \beta \in \mathcal{N}$ with $\alpha \subset \beta$ we have a commutative diagram

$$\begin{array}{ccc} \mathcal{O}_{\widetilde{W}_\alpha} & \xrightarrow{r_{\beta\alpha}} & (p_{\alpha\beta})_*(\mathcal{O}_{\widetilde{W}_\beta}) \\ {\scriptstyle F_\alpha^*}\downarrow & & \downarrow{\scriptstyle (p_{\alpha\beta})_*(F_\beta^*)} \\ \mathcal{O}_{\widetilde{W}_\alpha} & \xrightarrow{r_{\beta\alpha}} & (p_{\alpha\beta})_*(\mathcal{O}_{\widetilde{W}_\beta}) \end{array}$$

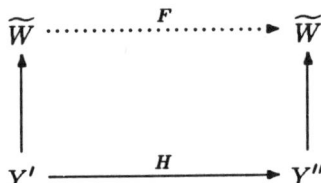

Here $r_{\beta\alpha}$ is the structure map (it is obtained by extending the corresponding structure map for $\mathcal{O}_{W_\bullet}$ to be A–linear). We put $\psi_\alpha = G^*_\alpha$, $\alpha \in \mathcal{N}$. We also put $Y = W \amalg_X X'$ and $Y'' = W \amalg_X X''$ and $H = id \amalg f$. Our problem is to solve the extension problem

$$\begin{array}{ccc} \widetilde{W} & \cdots\cdots\cdots\blacktriangleright & \widetilde{W} \\ \uparrow & & \uparrow \\ Y' & \xrightarrow{\quad H \quad} & Y'' \end{array}$$

which we have already solved over each simplex.

We first reduce the above extension problem to a succession of extension problems of maps of subspaces defined by ideal sheaves with square zero. We may assume $\mathfrak{m}^n = 0$. We put $A_k = A/\mathfrak{m}^k$ and $W_k = W \times \operatorname{Spec}(A_k)$. We define X'_k (resp. X''_k) to be the analytic space with underlying space equal to the underlying space of X and structure sheaf $\mathcal{O}_{X'}/\mathfrak{m}^k\mathcal{O}_{X'}$ (resp. $\mathcal{O}_{X''}/\mathfrak{m}^k\mathcal{O}_{X''}$). We will solve successively the extension problems

$$\begin{array}{ccc} W_k & \overset{F_k}{\cdots\cdots\cdots\blacktriangleright} & W_k \\ \uparrow & & \uparrow \\ X'_k & \xrightarrow{\quad f_k \quad} & X''_k \end{array}$$

starting with the square ($k = 1$)

$$\begin{array}{ccc} W & \xrightarrow{\quad id \quad} & W \\ \uparrow & & \uparrow \\ X & \xrightarrow{\quad id \quad} & X \end{array}$$

Here f_k is the restriction of f to X'_k.

We make the induction hypothesis that we have constructed local solutions G_α, $\alpha \in \mathcal{N}$, which patch together to give a global solution $F_k \bmod \mathfrak{m}^k$. We define analytic subspaces Y'_{k+1} and Y''_{k+1} of W_{k+1} by $Y'_{k+1} = W_k \amalg_{X'_k} X'_{k+1}$ and $Y''_{k+1} = W_k \amalg_{X'_k} X''_{k+1}$. An easy extension of the argument above shows that Y'_{k+1} and Y''_{k+1} are closed subspaces of $\widetilde{W}$. We put $H_{k+1} = F_k \amalg f_{k+1}$ and set $\varphi = H^*_{k+1}$. Thus the $(k+1)$ extension problem will be solved if we can solve the extension problem

$$\begin{array}{ccc} W_{k+1} & \cdots\cdots\cdots\blacktriangleright & W_{k+1} \\ \uparrow & & \uparrow \\ Y'_{k+1} & \xrightarrow{\quad H_{k+1} \quad} & Y''_{k+1} \end{array}$$

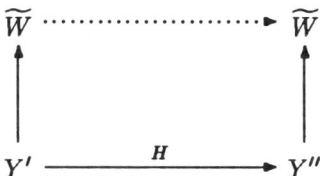

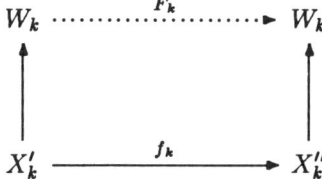

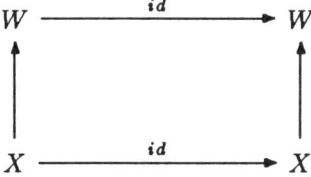

We claim that the ideal sheaf J'_{k+1} (resp. J''_{k+1}) of Y'_{k+1} in W_{k+1} (resp. Y''_{k+1} in W_{k+1}) has square zero. Indeed it is clear that the kernel of $\pi' : \mathcal{O}_{W_{k+1}} \longrightarrow \mathcal{O}_{Y'_{k+1}}$ coincides with that of $\mathcal{O}_{W_{k+1}} \longrightarrow \mathcal{O}_{W_k} \times \mathcal{O}_{X'_{k+1}}$ which is manifestly $\mathfrak{m}^k\mathcal{O}_{W_{k+1}} \cap I'_{k+1}$ (here I'_{k+1} is the ideal sheaf of X'_{k+1} in W_{k+1}).

We now solve the above extension problem. We will abbreviate W_{k+1} to W' and drop the subscript $k+1$ on Y'_{k+1}, Y''_{k+1}. Let $\alpha, \beta \in \mathcal{N}$ with $\alpha \subset \beta$ and let $s'_{\beta\alpha} : \mathcal{O}_{Y'_\alpha} \longrightarrow (p_{\alpha\beta})_*\mathcal{O}_{Y'_\beta}$ and $s''_{\beta\alpha} : \mathcal{O}_{Y''_\alpha} \longrightarrow (p_{\alpha\beta})_*\mathcal{O}_{Y''_\beta}$ be the structure maps. Since H_{k+1} is globally defined we have $s'_{\beta\alpha} \circ \varphi_\alpha = (p_{\alpha\beta})_*(\varphi_\beta) \circ s''_{\beta\alpha}$. We let $\delta_{\beta\alpha} \in \operatorname{Hom}_A(\mathcal{O}_{W'_\alpha}, (p_{\alpha\beta})_*\mathcal{O}_{W'_\beta})$ be defined by

$$\delta_{\beta\alpha} = (p_{\alpha\beta})_*(\psi_\beta) \circ r_{\beta\alpha} - r_{\beta\alpha} \circ \psi_\alpha .$$

Thus $\delta = \{\delta_{\beta\alpha}\}$ is a one-coboundary in the complex $C^\cdot(\operatorname{sd}\mathcal{N}, \operatorname{Hom}_A(\mathcal{O}_{W'}, \mathcal{O}_{W'}))$ that measures the failure of ψ to be a globally-defined homomorphism of simplicial sheaves.

We will now show that in fact δ is a one-cocycle in $C^\cdot(\operatorname{sd}\mathcal{N}, \operatorname{Der}(\mathcal{O}_W, J'))$. We first check that $\delta_{\alpha\beta}$ takes values in $(p_{\alpha\beta})_* J'_\beta$. Indeed

$$\begin{aligned}
(p_{\alpha\beta})_*(\pi'_\beta) \circ (p_{\alpha\beta})_*(\psi_\beta) \circ r_{\beta\alpha} &= (p_{\alpha\beta})_*(\pi'_\beta \circ \psi_\beta) \circ r_{\beta\alpha} \\
(p_{\alpha\beta})_*(\varphi_\beta \circ \pi''_\beta) \circ r_{\beta\alpha} &= (p_{\alpha\beta})_*(\varphi_\beta) \circ (p_{\alpha\beta})_*(\pi''_\beta) \circ r_{\beta\alpha} \\
&= (p_{\alpha\beta})_*(\varphi_\beta) \circ s''_{\beta\alpha} \circ \pi''_\alpha .
\end{aligned}$$

(The equation $(p_{\alpha\beta})_*(\pi''_\beta) \circ r_{\beta\alpha} = s''_{\alpha\beta} \circ \pi''_\alpha$ holds because $\pi'' : \mathcal{O}_{W'} \longrightarrow \mathcal{O}_{Y''}$ is globally defined.) Also,

$$(p_{\alpha\beta})_*(\pi'_\beta) \circ r_{\beta\alpha} \circ \psi_\alpha = s'_{\alpha\beta} \circ \pi'_\alpha \circ \psi_\alpha = s'_{\beta\alpha} \circ \varphi_\alpha \circ \pi''_\alpha .$$

We obtain

$$(p_{\alpha\beta})_*(\pi'_\beta) \circ \delta_{\beta\alpha} = \left((p_{\alpha\beta})_*(\varphi_\beta) \circ s''_{\beta\alpha} - s'_{\beta\alpha} \circ \varphi_\alpha\right) \circ \pi''_\alpha = 0.$$

The claim follows from the exact sequence $J'_\beta \longrightarrow \mathcal{O}_{W'_\beta} \longrightarrow \mathcal{O}_{Y'_\beta}$ and the exactness of $(p_{\alpha\beta})_*$. Hence δ is the image of a one-cocycle in $C^\cdot(\operatorname{sd}\mathcal{N}, \operatorname{Hom}_A(\mathcal{O}_{W'}, J'))$ which we continue to denote δ. But J' has square zero and $(p_{\alpha\beta})_*(\psi_\beta) \circ r_{\beta\alpha}$ and $r_{\beta\alpha} \circ \psi_\alpha$ are homomorphism of sheaves of A_{k+1}–algebras which coincide modulo $(p_{\alpha\beta})_*(J'_\beta)$. Hence $\delta_{\alpha\beta}$ is an A_{k+1}–derivation and $\delta \in C^\cdot(\operatorname{sd}\mathcal{N}, \operatorname{Der}(\mathcal{O}_W, J')) = C^\cdot\left(\operatorname{sd}\mathcal{N}, \operatorname{Hom}(\Omega^1_W, J')\right)$ as required. Since Ω^1_W is projective, by Theorem 4.3 we have

$$H^1\left(\operatorname{sd}\mathcal{N}, \operatorname{Hom}(\Omega^1_W, J')\right) = 0.$$

Hence there exists $\{D_\alpha\} \in C^0(\operatorname{sd}\mathcal{N}, \operatorname{Der}(\mathcal{O}_W, J'))$ such that $\delta_{\beta\alpha} = (p_{\alpha\beta})_*(D_\beta) \circ r_{\beta\alpha} - r_{\beta\alpha} \circ D_\alpha$.

We now resume the subscript $k+1$ notation and define $(F^*_{k+1})_\alpha = \psi_\alpha - D_\alpha$. To justify this definition we observe that since $\mathfrak{m}J' = 0$ in $\mathcal{O}_{W_{k+1}}$ we have $\psi(g)h = gh$ for $h \in J'$ and hence $\psi - D$ is a ring homomorphism. By construction F_{k+1} is globally defined. Moreover $\pi'_\alpha \circ F^*_{k+1} = \pi'_\alpha \circ \psi_\alpha - \pi'_\alpha \circ D_\alpha = \pi'_\alpha \circ \psi_\alpha = \varphi_\alpha \circ \pi''_\alpha$. Hence $F_{k+1}|Y'_{k+1} = H_{k+1}$. Thus we have solved the $(k+1)$–st extension problem.

Finally in order to continue the induction we need to extend our local map $F_\alpha = (F_{k+1})_\alpha : (W_{k+1})_\alpha \longrightarrow (W_{k+1})_\alpha$ to a map $G_\alpha : \widetilde{W}_\alpha \longrightarrow \widetilde{W}_\alpha$ such that $G|W_\alpha = id$

and $G|X'_\alpha = f'_\alpha$. We observe that F_α satisfies $F_\alpha|W_\alpha = id$ and $F_\alpha|(X'_{k+1})_\alpha = f'_\alpha|(X'_{k+1})_\alpha$. Hence F_α and f'_α glue together to give a map

$$K_\alpha : (W_{k+1})_\alpha \amalg_{(X'_{k+1})_\alpha} X'_\alpha \longrightarrow (W_{k+1})_\alpha \amalg_{(X''_{k+1})_\alpha} X''_\alpha.$$

We extend K_α to G_α using Cartan's Theorem B and the remark above as in the local case treated in the beginning of this proof. □

We next define the cofibered groupoid $\mathrm{Def}(R;\cdot)$ of deformations with infinitesimal base of the (fixed) resolvent $\mathfrak{g} = (X_*, W_*, R)$ for X. Let $A \in \mathrm{Obj}\ \mathcal{A}$ with maximal ideal $\mathfrak{m}$. An object of $\mathrm{Def}(R;A)$ is a sheaf R' of free differential graded $\mathcal{O}_{W_*} \otimes A$–algebras on W_* and a morphism $\rho : R' \longrightarrow R$ of $\mathcal{O}_{W_*} \otimes A$ algebras such that the following diagram is cocartesian

$$\begin{array}{ccc} R' & \longrightarrow & R \\ \uparrow & & \uparrow \\ A & \longrightarrow & \mathbb{C} \end{array}$$

Thus R'_α is a deformation of R_α for each simplex α and the structure maps $r'_{\alpha\beta} : p^*_{\alpha\beta} R'_\alpha \longrightarrow R'_\beta$ are deformations of $r_{\alpha\beta} : p^*_{\alpha\beta} R_\alpha \longrightarrow R_\beta$ for $\alpha \subset \beta$.

A morphism φ of the groupoid $\mathrm{Def}(R;A)$ is a homomorphism of sheaves of differential graded A–algebras such that $\varphi \equiv id_R \bmod \mathfrak{m}$. By Nakayama's Lemma we find that every morphism in $\mathrm{Def}(R;A)$ is an isomorphism. We now show that if $R' \in \mathrm{Obj}\ \mathrm{Def}(R;A)$ then the underlying algebra deformation is trivial; that is we may assume $R' = R \otimes A$ as a simplicial algebra (but not as a complex). □

Lemma 6.4. *Let R' be an object of* $\mathrm{Def}(R;A)$. *Then there is an isomorphism $\gamma : R \otimes A \longrightarrow R'$ of $\mathcal{O}_{W_*} \otimes A$–algebras.*

Proof. The main point is to construct a section $\sigma : R \longrightarrow R'$ of the reduction map $\rho : R' \longrightarrow R$. By definition $R = S_{\mathcal{O}_{W_*}}(M_*)$ with M_* a free $\mathcal{O}_{W_*}$–module. We let $\iota : M_* \longrightarrow R$ be the inclusion. Since $\rho : R' \longrightarrow R$ is surjective we may lift ι to $\iota' : M_* \longrightarrow R'$ with $\rho \circ \iota' = \iota$. We obtain an algebra map $\sigma : R \longrightarrow R'$ with $\rho \circ \sigma = id$.

Finally we define $\gamma : R \otimes A \longrightarrow R'$ to be the map satisfying $\gamma(r \otimes a) = a\sigma(r)$. Clearly γ reduces to id_R modulo $\mathfrak{m}$ and consequently γ is an isomorphism since the stalk R'_x is a flat A–module, $x \in W_*$. □

Remark. According to Lemma 6.4 we can assume that any deformation R' of R over A has $R \otimes A$ as underlying algebra. We make this assumption henceforth.

We now define a natural transformation $p : \mathcal{C}(L_X;\cdot) \longrightarrow \mathrm{Def}(R;\cdot)$ by "perturbing ∂". Let $\eta \in \mathrm{Obj}\ \mathcal{C}(L;A)$. Then $p(\eta)$ is defined to be the differential graded algebra R' with underlying algebra $R \otimes A$ and differential $\beta(\partial) + \beta(\eta)$ where β :

$\mathrm{Hom}_{\mathbb{C}}(R, R) \otimes A \longrightarrow \mathrm{Hom}_A(R \otimes A, R \otimes A)$ is the natural map (see the previous chapter). Let $\exp(\lambda) \in \mathrm{Mor}\ \mathcal{C}(L; A)$. Then we define

$$p(\exp(\lambda)) = \beta(\exp(\lambda))\,.$$

The proof of the following lemma is analogous to that of Lemma 5.4.

Lemma 6.5. *p is an equivalence of groupoids.* □

We now wish to construct a natural transformation $h : \mathrm{Def}(R; \cdot) \longrightarrow \mathrm{Def}(X; \cdot)$. Let $A \in \mathrm{Obj}\ \mathcal{A}$ with maximal ideal $\mathfrak{m}$. Let $R' \in \mathrm{Def}(R; A)$. Let $\alpha, \beta \in \mathcal{N}$ with $\alpha \subset \beta$ and $r'_{\beta\alpha} : p^*_{\alpha\beta}(R'_\alpha)^i \longrightarrow (R'_\beta)^i$ be the structure map. We have a commutative diagram

$$\begin{array}{ccc}
p^*_{\alpha\beta}(R'_\alpha)^{i-1} & \longrightarrow & (R'_\beta)^{i-1} \\
\Big\downarrow{\scriptstyle p^*_{\alpha\beta}\partial'_\alpha} & & \Big\downarrow{\scriptstyle \partial'_\beta} \\
p^*_{\alpha\beta}(R'_\alpha)^{i} & \longrightarrow & (R'_\beta)^{i}
\end{array}$$

Since $p^*_{\alpha\beta}$ is exact we have $\ker p^*_{\alpha\beta}\partial'_\alpha = p^*_{\alpha\beta}(\ker \partial'_\alpha)$ and $\mathrm{im}\, p^*_{\alpha\beta}\partial_\alpha = p^*_{\alpha\beta}(\mathrm{im}\, \partial'_\alpha)$. Thus the structure map $r'_{\beta\alpha}$ induces a map $H^i(r'_{\beta\alpha}) : p^*_{\alpha\beta}H^i(R'_\alpha) \longrightarrow H^i(R'_\beta)$ and we obtain an $\mathcal{O}_{W_*} \otimes A$–module $H^i(R')_*$ given by $H^i(R')_\alpha = H^i(R'_\alpha)$ where the H^i's denote cohomology sheaves. We next observe that these modules are supported on the image of $i : X_* \longrightarrow W_*$.

Lemma 6.6. *The sheaves $H^\cdot(R')_*$ are supported on the image of $i : X_* \longrightarrow W_*$. Moreover $H^i(R')_* = 0$, $i < 0$, $H^0(R')_*$ is a flat A-module and the natural map $H^0(R') \otimes_A \mathbb{C} \longrightarrow H^0(R) = i_*\mathcal{O}_{X_*}$ is an isomorphism.*

Proof. The fact that $H^i(R') = 0$, $i < 0$, follows from Lemma 5.3 applied to the stalks of R'. It then follows as in Chapter 5 that $H^0(R')$ is a flat A–module and that the natural map $H^0(R') \otimes_A \mathbb{C} \longrightarrow H^0(R)$ is an isomorphism. It remains to prove that $H^0(R')$ is supported on the image of i. Let $x \in W_*$ and $x \notin \mathrm{image}\, i$. Then $H^0(R_x) = 0$ whence $H^0(R')_x \otimes_A \mathbb{C} = 0$. The lemma now follows from Nakayama's Lemma. □

Corollary. *$i^{-1}H^0(R')$ is a sheaf of flat A–algebras.* □

We put $B'_* = i^{-1}H^0(R')$. We let k^* denote the natural restriction functor from sheaves on X to simplicial sheaves on U_*, recall $U_i = X^0_i$, $i \in I$. We let B''_* be the simplicial sheaf on U_* given by $B''_\alpha = B'_\alpha | U_\alpha$.

Lemma 6.7. *There exists a sheaf B' on X and an isomorphism $\eta_* : k^*B' \longrightarrow B''_*$ of simplicial sheaves of analytic A–algebras on U_*.*

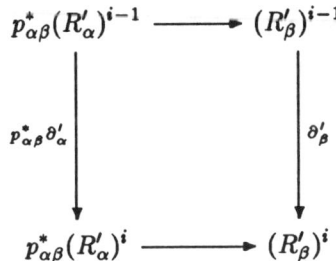

Proof. Since the reduction of B''_* modulo $\mathfrak{m}$ is $\mathcal{O}_{U_*}$ we find that the structure maps of B''_* are isomorphisms modulo $\mathfrak{m}$. Since the stalks of B''_* are flat A-modules it follows that the structure maps of B''_* are isomorphisms. The lemma now follows from Lemma 6.1. □

Unfortunately we need to check that the simplicial sheaf B'_* on X_* comes from a sheaf on X, that is that $j^*B' = B'_*$. This is intuitively clear but painful to make explicit.

Lemma 6.8. *The isomorphism η_* induces an isomorphism $\gamma_* : j^*B' \longrightarrow i^{-1}H^0(R')$ as sheaves of analytic A-algebras on X_*.*

Proof. We will abbreviate j^*B' to B and $i^{-1}H^0(R')$ to H for the proof of this lemma. We first construct an isomorphism $\gamma_\alpha : B|X_\alpha \longrightarrow H_\alpha$. Choose $U_i \in \mathcal{U}$ such that $U_i \cap X_\alpha \neq \emptyset$ (note that i might not be a vertex of α but there is a simplex $\alpha i \in \mathcal{N}$). We have isomorphisms $\eta_i : B|U_i \longrightarrow H_i|U_i$, $\mu_{\alpha i,i} : H_i|X_i \cap X_\alpha \longrightarrow H_{\alpha i}$ and $\mu_{\alpha i,\alpha} : H_\alpha|X_\alpha \cap X_i \longrightarrow H_{\alpha i}$. We define $\gamma_\alpha^{(i)} : B|X_\alpha \cap U_i \longrightarrow H_\alpha|X_\alpha \cap U_i$ by $\gamma_\alpha^{(i)} = (\mu_{\alpha i,\alpha}|X_\alpha \cap U_i)^{-1} \circ (\mu_{\alpha i,i}|X_\alpha \cap U_i) \circ \eta_i|X_\alpha \cap U_i$. We now prove that $\gamma_\alpha^{(i)} = \gamma_\alpha^{(j)}$ on $X_\alpha \cap U_i \cap U_j$ whenever $j \in I$ is such that the above intersection is not empty. We have on $U_i \cap U_j$ (see Lemma 6.1) $\eta_{ij} = \mu_{ij,i}\eta_i$ whence on $X_\alpha \cap U_i \cap U_j$ we have

$$\begin{aligned}
\gamma_\alpha^{(i)} &= \mu_{\alpha i,\alpha}^{-1} \circ \mu_{\alpha i,i} \circ \mu_{ij,i}^{-1} \circ \eta_{ij} \\
&= \mu_{\alpha i,\alpha}^{-1} \circ \mu_{\alpha ij,\alpha i}^{-1} \circ (\mu_{\alpha ij,\alpha i} \circ \mu_{\alpha i,i}) \circ \mu_{ij,i}^{-1} \circ \eta_{ij} \\
&= (\mu_{\alpha ij,\alpha i} \circ \mu_{\alpha i,\alpha})^{-1} \circ (\mu_{\alpha ij,ij} \circ \mu_{ij,i}) \circ \mu_{ij,i}^{-1} \circ \eta_{ij} \\
&= \mu_{\alpha ij,\alpha}^{-1} \circ \mu_{\alpha ij,ij} \circ \eta_{ij} \\
&= \gamma_\alpha^{(j)}.
\end{aligned}$$

It remains to check the compatibility conditions for the family $\{\gamma_\alpha\}$. Let $\alpha \subset \beta$ be given. We claim that $\gamma_\beta = \mu_{\beta\alpha} \circ \gamma_\alpha$ on X_β. To prove this choose U_i such that $U_i \cap X_\beta \neq \emptyset$. It then suffices to prove that $\gamma_\alpha^{(i)} = \mu_{\beta\alpha}^{-1}\gamma_\beta^{(i)}$ on $U_i \cap X_\beta$. Note that we have simplices αi and βi in $\mathcal{N}$ with $\alpha i \subset \beta i$. We have

$$\begin{aligned}
\mu_{\beta\alpha}^{-1}\gamma_\beta^{(i)} &= \mu_{\beta\alpha}^{-1} \circ \mu_{\beta i,\beta}^{-1} \circ \mu_{\beta i,i} \circ \eta_i \\
&= \mu_{\alpha i,\alpha}^{-1} \circ \mu_{\beta i,\alpha i}^{-1} \circ \mu_{\beta i,i} \circ \eta_i \\
&= \mu_{\alpha i,\alpha}^{-1} \circ \mu_{\alpha i,i} \circ \eta_i \\
&= \gamma_\alpha^{(i)}.
\end{aligned}$$

The lemma follows. □

We define h on objects by

$$h(R') = B'$$

and on morphisms we define $h(\varphi)$ to satisfy

$$k^*h(\varphi) = i^{-1}H_0(\varphi).$$

Before proving that h is full and surjective on isomorphism classes, we need to study certain functors of [F], page 64. Let R_α be a differential graded algebra over

$\mathcal{O}_{W_\alpha}$. We define a differential graded algebra $q_\alpha^* R_\alpha$ by the formula

$$(q_\alpha^* R_\alpha)_\beta = \begin{cases} p_{\alpha\beta}^* R_\alpha & \text{if } \alpha \subset \beta \\ \mathcal{O}_{W_\beta} & \text{otherwise.} \end{cases}$$

It is easy to check that if S is a differential graded algebra over $\mathcal{O}_{W_*}$ then

$$\mathrm{Hom}_{\mathcal{O}_{W_*}}(q_\alpha^* R_\alpha, S) = \mathrm{Hom}_{\mathcal{O}_{W_\alpha}}(R_\alpha, S_\alpha).$$

Here the Hom's denote homomorphisms of differential graded algebras. We note the formula that if F_α is a free $\mathcal{O}_{W_\alpha}$–module then (here $S_{\mathcal{O}_{W_\alpha}}(F_\alpha)$ means the symmetric algebra on F_α)

$$q_\alpha^* S_{\mathcal{O}_{W_\alpha}}(F_\alpha) = S_{\mathcal{O}_{W_*}}(p_\alpha^* F_\alpha).$$

Remark. Let (M, ∂) be a differential graded $\mathcal{O}_{W_*}$ module with underlying module M given by $M = \bigoplus_\alpha p_\alpha^* F_\alpha$. Let $i_\alpha : p_\alpha^* F_\alpha \longrightarrow M$ and $\pi_\alpha : M \longrightarrow p_\alpha^* F_\alpha$ be the inclusion and projection. Put $\partial_{\beta\alpha} = \pi_\beta \circ \partial \circ i_\alpha$. Then

$$\partial_{\beta\alpha} \in \mathrm{Hom}_{\mathcal{O}_{W_*}}(p_\alpha^* F_\alpha, p_\beta^* F_\beta) = \mathrm{Hom}_{\mathcal{O}_{W_\alpha}}\left(F_\alpha, (p_\beta^* F_\beta)_\alpha\right).$$

Since $(p_\beta^* F_\beta)_a = 0$ unless $\alpha \supseteq \beta$ we see that $\partial_{\beta\alpha} = 0$ unless $\alpha \supseteq \beta$. Thus if $\mathcal{N}' \subset \mathcal{N}$ is a subcomplex then the submodule $M' \subset M$ given by $M' = \bigoplus_{\alpha \in \mathcal{N}'} p_\alpha^* F_\alpha$ satisfies $\partial M' \subset M'$.

In what follows we will have to consider simultaneously the functors q_α^* for the simplicial space $\widetilde{W}_* = W_* \times \operatorname{Spec} A$, $A \in \operatorname{Obj} \mathcal{A}$ and the simplicial subspace W_*. In order to prevent confusion if $\alpha \in \mathcal{N}$ we will use $\tilde{q}_\alpha^*$ for the functors just defined for the case of $\widetilde{W}_*$ and q_α^* for the case of W_*. We note that $\tilde{q}_\alpha^*$ is a deformation of q_α^* in the sense that if R'_α is a deformation of R_α then $\tilde{q}_\alpha^* R'_\alpha$ is a deformation of $q_\alpha^* R_\alpha$.

We will need the following elementary lemma. Let A_* be an $\mathcal{O}_{W_*}$–algebra and F_α be an $\mathcal{O}_{W_\alpha}$–module. We define an A_α–algebra R_α by $R_\alpha = S_{A_\alpha}(F_\alpha \otimes_{\mathcal{O}_{W_\alpha}} A_\alpha)$, here S_{A_α} denotes the symmetric algebra.

Lemma 6.9. $S_{A_*}(p_\alpha^* F_\alpha \otimes_{\mathcal{O}_{W_\alpha}} A_*) = q_\alpha^* R_\alpha \otimes_{q^* \alpha A_\alpha} A_*$.

Proof.

$$\begin{aligned} S_{A_*}(p_\alpha^* F_\alpha \otimes_{\mathcal{O}_{W_\alpha}} A_*) &= S_{\mathcal{O}_{W_*}}(p_\alpha^* F_\alpha) \otimes_{\mathcal{O}_{W_*}} A_* \\ &= q_\alpha^* S_{\mathcal{O}_{W_\alpha}}(F_\alpha) \otimes_{\mathcal{O}_{W_*}} A_* \\ &= q_\alpha^* S_{\mathcal{O}_{W_\alpha}}(F_\alpha) \otimes_{\mathcal{O}_{W_*}} q_\alpha^* A_\alpha \otimes_{q_\alpha^* A_\alpha} A_* \\ &= q_\alpha^* \left(S_{\mathcal{O}_{W_\alpha}}(F_\alpha) \otimes_{\mathcal{O}_{W_\alpha}} A_\alpha\right) \otimes_{q_\alpha^* A_\alpha} A_* \\ &= q_\alpha^* R_\alpha \otimes_{q_\alpha^* A_\alpha} A_*. \end{aligned}$$

□

Corollary. *Let M be a free $\mathcal{O}_{W_*}$–module and F_α and A_* be as above. Then*

$$S_{A_*}\Big((p_\alpha^* F_\alpha \otimes A_*) \oplus (M \otimes A_*)\Big) = S_{A_*}(M \otimes A_*) \otimes_{A_*} (q_\alpha^* R_\alpha \otimes_{q_\alpha^* A_\alpha} A_*).$$ □

Remark. We record two more properties of the functor S which we will need in the proof of Lemma 6.10.

(i) $S_{A_*}(V \oplus W) = S_{S_{A_*}(V)}(W \otimes_{A_*} S_{A_*}(V))$

and

(ii) $S_{B_*}(V \otimes_{A_*} B_*) = S_{A_*}(V) \otimes_{A_*} B_*$.

We now construct a canonical increasing filtration $\{R(n)\}_{n \geq -1}$ of a simplicial differential graded $\mathcal{O}_{W_*}$-algebra R such that the underlying simplicial graded algebra is free.

Lemma 6.10. *Let R be a simplicial differential graded $\mathcal{O}_{W_*}$-algebra such that the underlying simplicial graded algebra is free. Choose free $\mathcal{O}_{W_\alpha}$-algebras F_α, $\alpha \in \mathcal{N}$, such that $R = S_{\mathcal{O}_{W_*}}\left(\bigoplus_\alpha p_\alpha^* F_\alpha\right)$ as a simplicial graded algebra. Define $R(n)$, a sub differential graded algebra, by $R(-1) = \mathcal{O}_{W_*}$ and $R(n) = S_{\mathcal{O}_{W_*}}\left(\bigoplus_{|\alpha| \leq n} p_\alpha^* F_\alpha\right)$, $n \geq 0$. (Note $R(n)$ is taken into itself by the differential of R by the above remark.) We have*

(i) *R is free over $R(n)$ for all n and $R(n)$ is free over $R(m)$, all $n > m$.*
(ii) *If α is a simplex of N of dimension n then the induced filtration of differential graded algebras*
$$\mathcal{O}_{W_\alpha} \subset R(0)_\alpha \subset R(1)_\alpha \subset \cdots \subset R_\alpha$$
satisfies $R(n)_\alpha = R(n+1)_\alpha = \cdots = R_\alpha$.
(iii) *We have the following formula for passing from $R(n)$ to $R(n+1)$ as differential graded algebras*
$$R(n+1) = \bigotimes_{|\beta|=n+1} \left(q_\beta^* R_\beta \otimes_{q_\beta^* R(n)_\beta} R(n)\right)$$
where the tensor product is taken over the simplicial differential graded algebra $R(n)$.

Proof. We prove only the formula (iii). Put $V = \bigotimes_{|\alpha| \leq n} p_\alpha^* F_\alpha$ whence $R(n) = S_{\mathcal{O}_{W_*}}(V)$. Then

$$\begin{aligned} R(n+1) &= S_{\mathcal{O}_{W_*}}\left(V \oplus \bigoplus_{|\beta|=n+1} p_\beta^* F_\beta\right) \\ &= S_{R(n)}\left(\left(\bigoplus_{|\beta|=n+1} p_\beta^* F_\beta\right) \otimes_{\mathcal{O}_{W_*}} R(n)\right). \end{aligned}$$

We now apply the corollary to Lemma 6.9 one simplex at a time to pull the $p_\beta^* F_\beta$'s outside the tensor product and obtain formula (iii). The reader will check using the adjoint formula for q_β^* that the differential ∂ on $R(n+1)$ coincides with the tensor product differential. That is, we have $\partial = \sum_{|\beta|=n+1} \Big(q_\beta^*(\partial_\beta) \otimes 1 + 1 \otimes \partial_n\Big) \otimes 1_\beta$ where ∂_β is the value of ∂ on $R(n+1)_\beta = R_\beta$, $\partial_n = \partial | R(n)$ and 1_β is the tensor product of identity maps over all γ, $|\gamma| = n+1$, $\gamma \neq \beta$. □

Remark. The formula (iii) shows that the above filtration is canonical. We will call $R(n)$ the n–skeleton of R.

We now return to the embedding $i : X_* \longrightarrow W_*$ of simplicial analytic spaces considered above. Suppose R is a free simplicial differential graded algebra on W_* and $\varepsilon : R \longrightarrow i_*\mathcal{O}_{X_*}$ is a surjection. The next lemma gives the inductive formula used in [F], Theorem 2.6, to construct a resolvent for $i_*\mathcal{O}_{X_*}$. We leave the proof to the reader.

Lemma 6.11. *The free simplicial differential graded algebra R is a resolvent for $i_*\mathcal{O}_{X_*}$ if and only if for all $n \geq -1$ and every $(n+1)$–simplex β the homomorphism $\varepsilon_\beta : R_\beta \longrightarrow (i_*\mathcal{O}_{X_*})_\beta$ is a resolvent for $(i_*\mathcal{O}_{X_*})_\beta$ as an $R(n)_\beta$–algebra.* □

We can now prove that h is full and surjective on isomorphism classes.

Lemma 6.12. *The functor h is surjective on isomorphism classes.*

Proof. Let $B' \in \mathrm{Der}(B; A)$ be given. We construct $R' \in \mathrm{Def}(R; A)$ by induction on the n–skeleton of R. Precisely, we will construct inductively a sequence of free differential graded $\mathcal{O}_{W_*} \otimes_k A$–algebras $R'(0) \subset R'(1) \subset \cdots$ flatly deforming the canonical filtration of R by its skeleta. We recall from Lemma 6.2 that we have a surjection $\varepsilon' : \mathcal{O}_{W_*} \otimes A \longrightarrow i_*B'_*$ where $B'_* = j^*B'$. We put $R'(-1) = \mathcal{O}_{W_*} \otimes A$.

We first construct $R'(0)$. We have $R(0) = \bigotimes_{|\alpha|=0} q_\alpha^* R_\alpha$ where R_α is a resolvent for $(i_\alpha)_*\mathcal{O}_{X_\alpha}$ over $\mathcal{O}_{W_\alpha}$. We apply Lemma 5.8 with $S = \mathcal{O}_{W_\alpha}$, $S' = \mathcal{O}_{W_\alpha} \otimes A$, $B = i_*\mathcal{O}_{X_\alpha}$ and $B' = i_*B'_\alpha$ to obtain R'_α, a free $\mathcal{O}_{W_\alpha} \otimes A$–algebra deforming R_α and $\varepsilon'_\alpha : R'_\alpha \longrightarrow i_*B'_\alpha$. We define $R'(0)$ by $R'(0) = \bigotimes_{|\alpha|=0} \tilde{q}_\alpha^* R'_\alpha$ and ε' to be the tensor product of the maps $\tilde{q}_\alpha^*\varepsilon'_\alpha$ above. Then $\varepsilon' : R'(0) \longrightarrow i_*B'$ is a surjective quasi-isomorphism over the zero skeleton of $\mathcal{N}$.

We now assume inductivity that we have constructed $R'(n)$, a deformation of $R(n)$, and $\varepsilon' : R'(n) \longrightarrow i_*B'$ a surjective quasi-isomorphism over the n–skeleton of $\mathcal{N}$. Let β be an $(n+1)$–simplex. We apply Lemma 5.8 with $S = R(n)_\beta$, $S' = R'(n)_\beta$ and $R = R_\beta$ to obtain R'_β, a free $R'(n)_\beta$–algebra which is a deformation of R_β, and $\varepsilon'_\beta : R'_\beta \longrightarrow B'_\beta$, a surjective quasi-isomorphism of differential graded S'–algebras. We define $R'(n+1)$ by

$$R'(n+1) = \bigotimes_{|\beta|=n+1} \tilde{q}_\beta^* R'_\beta \otimes_{\tilde{q}_\beta^* R'(n)_\beta} R'(n)$$

and ε' to be the product of the $\tilde{q}_\beta^*\varepsilon'_\beta$. We leave to the reader the task of verifying that $R'(n+1)$ has the required properties. □

Lemma 6.13. *The functor h is full.*

Proof. Let $B', B'' \in \text{Obj } \text{Def}(X;A)$ and $R', R'' \in \text{Obj } \text{Def}(R;A)$ be given with $h(R') = B'$ and $h(R'') = B''$. Let $f : B' \longrightarrow B''$ be a morphism. We apply Lemma 6.3 to lift $i_* f$ to a morphism $F : R'(-1) = \mathcal{O}_{W_\bullet} \otimes A \longrightarrow R''(-1) = \mathcal{O}_{W_\bullet} \otimes A$. We will construct a lift of F to a morphism $g : R' \longrightarrow R''$ such that $g|R'(n+1)$ preserves the above tensor product decomposition by induction on the above filtration. By Lemma 6.4 we may assume $R' = R'' = \widetilde{R} = R \otimes_{\mathbb{k}} A$ as simplicial graded algebras. We let ∂' denote the differential in R' and ∂'' be the differential in R''. By induction we may assume g has been constructed on the sub simplicial differential graded algebra $S = R'(n)$ and we are required to extend g to a morphism $\tilde{g} : \left(\tilde{q}_\beta^* \widetilde{R}_\beta \otimes_{\tilde{q}_\beta^* S_\beta} S, \partial'\right) \longrightarrow \left(\tilde{q}_\beta^* \widetilde{R}_\beta \otimes_{\tilde{q}_\beta^* S_\beta} S, \partial''\right)$ for each β with $|\beta| = n+1$. By the adjunction formula for $\tilde{q}_\beta^*$ it suffices to construct a homomorphism $\tilde{g}_\beta : (\widetilde{R}_\beta, \partial'_\beta) \longrightarrow (\widetilde{R}_\beta, \partial''_\beta)$ which restricts to the given homomorphism $g_\beta : (S_\beta, \partial') \longrightarrow (S_\beta, \partial'')$ and reduces to the identity modulo $\mathfrak{m}$. The lemma now follows from Lemma 5.9. □

Lemma 6.14. *The functor* $h : \text{Def}(R;A) \longrightarrow \text{Def}(X;A)$ *induces an isomorphism* $\text{Iso } \text{Def}(R;A) \longrightarrow \text{Iso } \text{Def}(X;A)$.

Proof. We have seen that h is surjective on isomorphism classes and full. □

The proof of Theorem C now follows in the same way as that of Theorem A.

7. The Comparison of the Tangent Complex and the Kodaira-Spencer Algebra of a Complex Manifold

The purpose of this chapter is to prove the following theorem.

THEOREM D. *Let X be a complex manifold with tangent complex L_X. Let $\mathcal{L}(X)$ be the Kodaira-Spencer algebra of X. Then L_X and $\mathcal{L}(X)$ are quasi-isomorphic as differential graded Lie algebras.*

Remark. We recall that $\mathcal{L}(X)$ is defined by

$$\mathcal{L}^q(X) = \mathcal{A}^{0,q}\left(X, T^{1,0}(X)\right), \quad q \geq 0.$$

We obtain the following corollary of Theorem D and the comparison theorem.

Corollary. *There is an isomorphism of complete local $\mathbb{C}$-algebras*

$$R_{L_X} \cong R_{\mathcal{L}(X)}.$$ □

The proof of Theorem D will require some preparation. We begin with two easy lemmas.

Lemma 7.1. *Let A and B be differential graded algebras and $\varphi : A \longrightarrow B$ a homomorphism of differential graded algebras. Let $\varphi_* : \mathrm{Der}(A) \longrightarrow \mathrm{Der}(A,B)$ and $\varphi^* : \mathrm{Der}(B) \longrightarrow \mathrm{Der}(A,B)$ be the induced maps (see below). Then the fiber product F of $\mathrm{Der}(A)$ and $\mathrm{Der}(B)$ over $\mathrm{Der}(A,B)$ is a differential graded Lie subalgebra of $\mathrm{Der}(A) \times \mathrm{Der}(B)$.*

Proof. The maps φ_* and φ^* are defined by

$$\begin{aligned}(\varphi_* D)(a) &= \varphi(Da)\\ (\varphi^* \bar{D})(b) &= \bar{D}(\varphi(b)).\end{aligned}$$

We say that a pair $(D, \bar{D}) \in \mathrm{Der}(A) \times \mathrm{Der}(B)$ are φ–related (or compatible) if and only if

$$\varphi_* D = \varphi^* \bar{D}$$

or by the above

$$\varphi(Da) = \bar{D}\varphi(a), \qquad \text{for all } a \in A.$$

Thus $(D, \bar{D}) \in F$ if and only if D and $\bar{D}$ are φ–related. Thus to show F is a sub graded Lie algebra of $\mathrm{Der}(A) \times \mathrm{Der}(B)$ it suffices to show that if D_1 (resp. D_2) is φ–related to $\bar{D}_1$ (resp. $\bar{D}_2$) then the graded bracket $[D_1, D_2]$ is φ–related to the graded bracket $[\bar{D}_1, \bar{D}_2]$. Assume that $\deg D_1 = \deg \bar{D}_1 = p$ and $\deg D_2 = \deg \bar{D}_2 = q$ where

$$\begin{aligned}[D_1, D_2] &= D_1 \circ D_2 - (-1)^{pq} D_2 \circ D_1\\ [\bar{D}_1, \bar{D}_2] &= \bar{D}_1 \circ \bar{D}_2 - (-1)^{pq} \bar{D}_2 \circ \bar{D}_1.\end{aligned}$$

Let $a \in A$ be given. Then

$$\begin{aligned}\varphi_*[D_1, D_2]a &= \varphi(D_1 D_2 a - (-1)^{pq} D_2 D_1 a)\\ &= \bar{D}_1 \varphi(D_2 a) - (-1)^{pq} \bar{D}_2 \varphi(D_1 a)\\ &= \bar{D}_1 \bar{D}_2 \varphi(a) - (-1)^{pq} \bar{D}_2 D_1 \varphi(a)\\ &= \varphi^*[\bar{D}_1, \bar{D}_2]a.\end{aligned}$$

Finally we observe that since φ preserves differentials we have

$$\varphi \circ \partial_A = \partial_B \circ \varphi$$

where ∂_A is the differential of A and ∂_B is the differential of B. Thus $\partial_A \in \mathrm{Der}(A)$ and $\partial_B \in \mathrm{Der}(B)$ are φ–related, hence if $(D, \bar{D}) \in F$ then $([\partial_A, D], [\partial_B, \bar{D}]) \in F$ by the above. But the definition of the differential d in $\mathrm{Der}(A) \times \mathrm{Der}(B)$ is

$$d(D, \bar{D}) = ([\partial_A, D], [\partial_B, \bar{D}]).$$

With this the lemma is proved. □

Lemma 7.2. *Let A and B be differential graded algebras and $\varphi : A \longrightarrow B$ a homomorphism. Suppose the natural maps $\varphi_* : \mathrm{Der}(A, A) \longrightarrow \mathrm{Der}(A, B)$ and $\varphi^* : \mathrm{Der}(B, B) \longrightarrow \mathrm{Der}(A, B)$ are quasi-isomorphisms of complexes. Assume further that the map $\varphi_* - \varphi^*$ maps $\mathrm{Der}(A, A) \times \mathrm{Der}(B, B)$ onto $\mathrm{Der}(A, B)$. Then the* *projection maps π_1 and π_2 in the following fiber square are also quasi-isomorphisms*

$$\begin{array}{ccc} F & \xrightarrow{\pi_2} & \mathrm{Der}(B,B) \\ {\scriptstyle \pi_1}\downarrow & & \downarrow{\scriptstyle \varphi^*} \\ \mathrm{Der}(A,A) & \xrightarrow{\varphi_*} & \mathrm{Der}(A,B) \end{array}$$

Proof. It suffices to check that if F is the fiber product of two quasi-isomorphisms $p_1 : A \longrightarrow C$ and $p_2 : B \longrightarrow C$ of complexes with $p_1(A) + p_2(B) = C$ then the projections $\pi_1 : F \longrightarrow A$ and $\pi_2 : F \longrightarrow B$ are quasi-isomorphisms. We have a short exact sequence of complexes

$$0 \longrightarrow F \xrightarrow{j} A \times B \xrightarrow{\pi} C \longrightarrow 0$$

where $j(f) = (\pi_1(f), \pi_2(f))$ and $\pi(a,b) = p_1(a) - p_2(b)$. We obtain a long exact sequence of cohomology

$$\cdots \longrightarrow H^i(F) \longrightarrow H^i(A) \times H^i(B) \longrightarrow H^i(C) \longrightarrow H^{i+1}(F) \longrightarrow \cdots \quad .$$

But the maps $H^i(A) \times H^i(B) \longrightarrow H^i(C)$ are onto so the above long exact sequence decomposes into short exact sequences

$$0 \longrightarrow H^i(F) \longrightarrow H^i(A) \times H^i(B) \longrightarrow H^i(C) \longrightarrow 0$$

and consequently $H^i(F)$ is the fiber product of the isomorphisms $H^i(A) \longrightarrow H^i(C)$ and $H^i(B) \longrightarrow H^i(C)$, and the maps $H^\cdot(\pi_1)$ and $H^\cdot(\pi_2)$ are isomorphisms as well. □

We will say that a fiber square as above is *good* if the hypotheses of Lemma 7.2 are satisfied.

We recall that in Chapter 4 we defined a resolvent $\mathfrak{g} : (X_*, W_*, R_*)$ for any complex analytic space. In case X is smooth we can refine the construction as follows. Let $\mathcal{U} = \{X_i, : i \in I\}$ be a locally finite cover of X by *open* coordinate domains and $\varphi_i : X_i \longrightarrow W_i$, $i \in I$, be the associated coordinate charts with W_i a ball in $\mathbb{C}^n$, $n = \dim X$. We assume that X_i has compact closure $\bar{X}_i$, $i \in I$, in X and that φ_i extends to an isomorphism from a neighbourhood V_i of $\bar{X}_i$ to a neighbourhood of W_i, $i \in I$. We let $\mathcal{N} = \mathrm{Nerve}(\mathcal{U})$ and assume that the nerve of $\{V_i : i \in I\}$ is also $\mathcal{N}$. We now define simplicial schemes W_* and X_* as in Chapter 4 (here it is more convenient to work with the open sets, our approach now coincides with that of [P]). Thus $X_\alpha = X_{i_0} \cap \cdots \cap X_{i_m}$ and $W_\alpha = W_{i_0} \times \cdots \times W_{i_m}$ for $\alpha = (i_0, i_i, \ldots, i_m)$. We define $i_\alpha : X_\alpha \longrightarrow W_\alpha$ as follows. Let $\gamma_\alpha : X_\alpha \longrightarrow \prod_{j=0}^{m} X_\alpha$ be the diagonal embedding and $\iota_{j\alpha} : X_\alpha \longrightarrow X_{i_j}$, $0 \leq j \leq m$, be the inclusion. We define $\iota_\alpha = \prod_{j=0}^{m} \iota_{j\alpha}$ and $\varphi_\alpha = \prod_{j=0}^{m} \varphi_{i_j}$. We then define $i_\alpha = \varphi_\alpha \circ \iota_\alpha \circ \gamma_\alpha$.

We put $\Gamma_\alpha = i_\alpha(X_\alpha)$. We claim that Γ_α is a smooth complete intersection in W_α. Indeed let $\{z^a_{i_j},\ 1 \leq a \leq n\}$ be the coordinate functions on W_{i_j}. We will identify these functions with their pull-backs to W_α whenever i_j is a vertex of α.

Now let α be as above. For each pair of vertices i_j, i_k of α with $j < k$ we define functions $f^a_{i_j i_k}$, $1 \le a \le n$ by

$$f^a_{i_j i_k} = z^a_{i_j} - z^a_{i_k}, \quad 1 \le a \le n.$$

Then the ideal of Γ_α is generated by the nm functions $\left\{ f^a_{i_0 i_j} : 1 \le j \le m, 1 \le a \le n \right\}$. Clearly the Koszul complex corresponding to these generators is exact.

We will now give a construction of R_* designed to take advantage of the facts that the X_i's are smooth and $i_\alpha(X_\alpha) \subset W_\alpha$ is a smooth complete intersection, $\alpha \in \mathcal{N}$. From the latter observation one might expect that it would be possible to take R_α to be a Koszul resolution of $i_\alpha(X_\alpha)$. Unfortunately this is not correct for the requirement that R_* be free as an $\mathcal{O}_{W_*}$-algebra forces us to carry forward to R_β all the free generators from R_α, $\alpha \subset \beta$, and thus as soon as $|\beta| \ge 2$ one finds that R_β cannot be a minimal resolution. The point of the next lemma is to show that R_β can be obtained from a Koszul resolution in a very simple way. In particular R_β can be chosen to be *finitely generated* over $\mathcal{O}_{W_\beta}$ as a graded commutative algebra.

Definition. Let $K^\bullet$ be a sheaf of differential graded algebras over a topological space X and $(M^\bullet, \partial_{M^\bullet})$ a differential graded $K^\bullet$-module which is a free $K^\bullet$-module. Let $R^\bullet = S_{K^\bullet}(M^\bullet)$. In the case that $R^\bullet$ is a finitely generated algebra over $K^\bullet$, so $M^\bullet$ is of finite rank as a $K^\bullet$-module, we define $\deg_{K^\bullet}(R^\bullet)$, the degree of $R^\bullet$ over $K^\bullet$, to be the maximum of the absolute values of the degrees of the elements of a basis of $M^\bullet$. We will say that $R^\bullet$ is *contractible* if there exists a decomposition $M^\bullet = M^\bullet_1 \oplus M^\bullet_2$ such that ∂_M carries $M^\bullet_1$ isomorphically onto $M^\bullet_2$.

We will use the notation that if $R^\bullet$ is a differential graded algebra free over an algebra $\mathcal{O}$, then $R^{(-k)}$ denotes the subalgebra generated by all generators of degree greater than or equal to $-k$.

Lemma 7.3. *There exists a resolvent R_* for X such that R_γ is a finitely generated $\mathcal{O}_{W_\gamma}$-module and is a contractible extension of degree $|\gamma|$ of a Koszul resolution of $(i_\gamma)_*\mathcal{O}_{X_\gamma}$ over $\mathcal{O}_{W_\gamma}$ for each $\gamma \in \mathcal{N}$.*

Proof. Since $\varphi_i(X_i) = W_i$ we take $R_\gamma = \mathcal{O}_{W_\gamma}$ if $|\gamma| = 0$. For each $\gamma \in \mathcal{N}$, with $|\gamma| = 1$, we choose a set of generators f^a_γ, $1 \le a \le n$ for the ideal of X_γ such that the differentials of the f^a_γ are independent at each point of X_γ (this is possible by the discussion above). We then take R_γ to be the corresponding Koszul resolution of the ideal sheaf of X_γ over $\mathcal{O}_{W_\gamma}$. Thus we can assume $|\gamma| \ge 2$. We now construct R_γ by induction on $|\gamma|$.

We assume by induction that if $\gamma = (i_0, i_1, \dots, i_k)$ with $k < m$ then R_γ is a contractible extension of degree k of the Koszul resolution of $i_\gamma(\mathcal{O}_{X_\gamma})$ corresponding to the generators $\left\{ f^a_{i_0 i_j} : 1 \le a \le n \; 1 \le j \le k \right\}$ of the ideal sheaf of X_γ in W_γ. Now let γ be an m-simplex. If $\beta \subset \gamma$ then we have $R_\beta = S_{\mathcal{O}_{W_\beta}} \left(\bigoplus_{\mu=-|\beta|}^{-1} F^\mu_\beta \right)$ by our induction hypothesis. We define a differential graded $\mathcal{O}_{W_\gamma}$-algebra $A^\bullet_0$ by

$A_0^\bullet = \mathcal{S}o_{W_\gamma}\left(\bigoplus_{\beta\in\mathcal{B}}\bigoplus_{\mu=-m+1}^{-1} p_{\beta\gamma}^* F_\beta^\mu\right)$. Here $\mathcal{B}$ denotes the set of codimension one faces of γ. Thus $A_0^\bullet$ is the differential graded $\mathcal{O}_{W_\gamma}$–algebra freely generated by the pull-backs of the generators of $R_\beta^\bullet$ as β runs through the codimension one faces of γ. We observe that $H_0(A_0^\bullet) = (i_\gamma)_*\mathcal{O}_{X_\gamma}$ since the lifts of the generators of the ideals of $i_\beta(X_\beta)$, $\beta \in \mathcal{B}$, generate the ideal of $i_\gamma(X_\gamma)$, $|\gamma| \geq 2$. We have natural inclusion maps $r_{\gamma\beta} : p_{\beta\gamma}^* R_\beta^\bullet \longrightarrow A_0^\bullet$, $\beta \in \mathcal{B}$. Now let $I_\gamma^\bullet$ denote the differential ideal in $A_0^\bullet$ generated by $\left\{ r_{\gamma\beta'} \circ r_{\beta'\alpha}(x) - r_{\gamma\beta''} \circ r_{\beta''\alpha}(x) : x \in \bigoplus_{\mu=-m+2}^{-1} F_\alpha^\mu \right\}$ as α runs through the codimension two faces of γ and β' and β'' are the two distinct codimension one faces of γ containing α. We let $\bar{A}_0^\bullet$ be the quotient of $A_0^\bullet$ by $I_\gamma^\bullet$. Let $\alpha \subset \gamma$ be a proper face. We construct $r_{\gamma\alpha} : p_{\alpha\gamma}^* R_\alpha^\bullet \longrightarrow \bar{A}_0^\bullet$ as follows. Choose a codimension one face β of γ such that $\alpha \subset \beta$. Define $r_{\gamma\alpha} = r_{\gamma\beta} \circ p_{\beta\gamma}^*(r_{\beta\alpha})$. The point of quotienting by $I_\gamma^\bullet$ is to make $r_{\gamma\alpha}$ well-defined. We next note that $\bar{A}_0^\bullet$ is still a free algebra over $\mathcal{O}_{W_\gamma}$. Indeed it is freely generated by $r_{\gamma\beta}\left(\bigoplus_{\mu=-m+2}^{-1} p_{\alpha\gamma}^* F_\alpha^\mu\right)$ as α runs through the codimension two faces of γ together with the "new generators" coming from the codimension one faces. It is also clear that $H_0(\bar{A}_0^\bullet) = (i_\gamma)_*\mathcal{O}_{X_\gamma}$. We will now add a *finite number* of free generators of degree -2 or less to kill the other homology groups of $\bar{A}_0^\bullet$.

We now choose $\left\{ f_{i_0 i_j}^a : 1 \leq a \leq n,\ 1 \leq j \leq m \right\}$ as generators for the ideal sheaf of X_γ in W_γ (see above). Let $m_1 = nm$ and order the above generators as $\left\{ f_1^{(0)}, \dots, f_{m_1}^{(0)} \right\}$. We may choose a subset $\left\{ g_1^{(-1)}, \dots, g_{m_1}^{(-1)} \right\}$ of the generators of $\bar{A}_0^\bullet$ of degree -1 such that $\partial g_j^{(-1)} = f_j^{(0)}$, $1 \leq j \leq m_1$. Then $B_0^\bullet = K_\gamma^\bullet = \mathcal{O}_{W_\gamma}\left[g_1^{(-1)}, \dots, g_{m_1}^{(-1)} \right]$ is a Koszul resolution of $(i_\gamma)_*\mathcal{O}_{X_\gamma}$. We may assume that the remainder of the generators $\left\{ f_1^{(-1)}, \dots, f_{n_1}^{(-1)} \right\}$ of $\bar{A}_0^\bullet$ of degree -1 are cycles (by subtracting off suitable linear combinations of $g_j^{(-1)}$, $1 \leq j \leq m_1$. We put $C_0^\bullet = B_0^\bullet\left[f_1^{(-1)}, \dots, f_{n_1}^{(-1)} \right]$ and define $B_1^\bullet$ by adjoining free generators $g_1^{(-2)}, \dots, g_{n_1}^{(-2)}$ to $C_0^\bullet$ and extending ∂ by the rule $\partial g_i^{(-2)} = f_i^{(-1)}$, $1 \leq i \leq n_1$. We note that $B_1^\bullet = \bar{A}_0^{(-1)}\left[g_1^{(-2)}, \dots, g_{n_1}^{(-2)} \right]$ and $B_1^\bullet$ is a contractible extension of $K_\gamma^\bullet$ whence $H_i(B_1^\bullet) = 0$, $i > 0$.

Now suppose we have constructed $B_k^\bullet$ for some k with $k \leq |\gamma| - 2$ with the properties that $B_k^\bullet = \bar{A}_0^{(-k)}\left[g_j^{(-i)} \right]$, $j \in J$, and $-i = \deg g_j^{(-i)}$ satisfies $-k-1 \leq -i \leq -2$ and $B_k^\bullet$ is a contractible extension of $K_\gamma^\bullet$ whence $H_i(B_k^\bullet) = 0$, $i > 0$. We put $\bar{A}_k^\bullet = \bar{A}_0^\bullet\left[g_j^{(-i)} \right]$. Let $f_1^{(-k-1)}, \dots, f_{n_{k+1}}^{(-k-1)}$ be the free generators of $\bar{A}_0^\bullet$ of degree $-k-1$. Let $C_k^\bullet$ be the differential graded subalgebra of $\bar{A}_k^\bullet$ given by $C_k^\bullet = B_k^\bullet\left[f_1^{(-k-1)}, \dots, f_{n_{k+1}}^{(-k-1)} \right]$. Then $C_k^\bullet$ is free over $B_k^\bullet$ on $f_1^{(-k-1)}, \dots, f_{n_{k+1}}^{(-k-1)}$. Since $H_k(B_k^\bullet) = 0$ and $B_k^\bullet \supset \bar{A}_0^{(-k)}$ we may assume $f_i^{(-k-1)}$ is a cycle, $1 \leq i \leq$

n_{k+1}. We then define $B^{\cdot}_{k+1} = C^{\cdot}_k\left[g_1^{(-k-2)}, \ldots, g_{n_{k+1}}^{(-k-2)}\right]$, the free commutative graded algebra on the indeterminates $g_i^{(-k-2)}$, $1 \le i \le n_{k+1}$, of degree $-k-2$. We extend ∂ to $B^{\cdot}_{k+1}$ by requiring $\partial g_i^{(-k-2)} = f_i^{(-k-1)}$, $1 \le i \le n_{k+1}$. Then $B^{\cdot}_{k+1} = \bar{A}_0^{(-k-1)}\left[g_j^{(-i)}; g_1^{(-k-2)}, \ldots, g_{n_{k+1}}^{(-k-2)}\right]$ and $B^{\cdot}_{k+1}$ is a contractible extension of $K^{\cdot}_{\gamma}$. We have completed the induction step.

We define $R^{\cdot}_{\gamma} = B^{\cdot}_{|\gamma|-1}$. Since $\bar{A}^{\cdot}_0 = \bar{A}_0^{(-|\gamma|+1)}$ we find that $R^{\cdot}_{\gamma}$ is a free algebra over $\bar{A}^{\cdot}_0$ of degree less than or equal to $|\gamma|$. The compositions $p^*_{\alpha\gamma}R^{\cdot}_{\alpha} \longrightarrow \bar{A}^{\cdot}_0 \longrightarrow R^{\cdot}_{\gamma}$ give the required structure maps $p^*_{\alpha\gamma}R^{\cdot}_{\alpha} \longrightarrow R^{\cdot}_{\gamma}$, all $\alpha \subset \gamma$. Continuing in this way we obtain R_* over all of $\mathcal{N}$.

□

We define a differential graded $\mathcal{O}_{X_\bullet}$-algebra $\mathcal{A}_{X_\bullet}$ by

$$\mathcal{A}_{X_\alpha} = (\mathcal{A}|X_\alpha, \bar{\partial})$$

where $\mathcal{A}$ denotes the sheafified Dolbeault algebra on X. We observe that $\mathcal{A}_{X_\bullet} = j^*(\mathcal{A})$ and that the inclusion $\iota : \mathcal{O}_{X_\bullet} \longrightarrow \mathcal{A}_{X_\bullet}$ is a quasi-isomorphism. The augmentation $\varepsilon : R_* \longrightarrow \bar{R}_* = i_*\mathcal{O}_{X_\bullet}$ is also a quasi-isomorphism. We put $\varphi = \iota \circ \varepsilon$ and obtain a quasi-isomorphism $\varphi : R_* \longrightarrow i_*\mathcal{A}_{X_\bullet}$. We obtain a fiber square

$$\begin{array}{ccc} F & \xrightarrow{\pi_2} & \mathrm{Der}(\mathcal{A}_{X_\bullet}, \mathcal{A}_{X_\bullet}) \\ \Big\downarrow{\scriptstyle \pi_1} & & \Big\downarrow{\scriptstyle \varphi^*} \\ \mathrm{Der}^+(R_*, R_*) & \xrightarrow{\varphi_*} & \mathrm{Der}^+(R_*, i_*\mathcal{A}_{X_\bullet}) \end{array}$$

with π_1 and π_2 homomorphisms of differential graded Lie algebras. To obtain the vertical arrow φ^* use the natural map $\mathrm{Der}(\mathcal{A}_{X_\bullet}, \mathcal{A}_{X_\bullet}) \longrightarrow \mathrm{Der}(i_*\mathcal{A}_{X_\bullet}, i_*\mathcal{A}_{X_\bullet})$.

Lemma 7.4. *Let W_* be a simplicial ringed space with underlying simplicial complex $\mathcal{N}$. Let $R^{\cdot}_*$ be a differential graded algebra over $\mathcal{O}_{W_\bullet}$, which is free as a graded algebra, $E^{\cdot}_*$ a free $R^{\cdot}_*$-module and $\varphi : M^{\cdot}_* \longrightarrow N^{\cdot}_*$ a quasi-isomorphism of differential graded R_*-modules such that M_* and N_* have locally acyclic homogeneous components. Then $\varphi_* : \mathrm{Hom}_{R^{\cdot}_*}(E^{\cdot}_*, M^{\cdot}_*) \longrightarrow \mathrm{Hom}_{R^{\cdot}_*}(E^{\cdot}_*, N^{\cdot}_*)$ is a quasi-isomorphism.*

Proof. We consider the induced map of double complexes

$$\varphi : C^{\cdot}\left(\mathrm{sd}\,\mathcal{N}, \mathrm{Hom}_{R_\bullet}(E^{\cdot}_*, M^{\cdot}_*)\right) \longrightarrow C^{\cdot}\left(\mathrm{sd}\,\mathcal{N}, \mathrm{Hom}_{R_\bullet}(E^{\cdot}_*, N^{\cdot}_*)\right).$$

By Theorem 4.3 (since $E^{\cdot}_*$ is free) the induced map on the cohomology of the total complexes is

$$H^{\cdot}(\varphi) : H^{\cdot}\left(\mathrm{Hom}_{R_\bullet}(E^{\cdot}_*, M^{\cdot}_*)\right) \longrightarrow H^{\cdot}\left(\mathrm{Hom}_{R_\bullet}(E^{\cdot}_*, N^{\cdot}_*)\right).$$

Thus it suffices to show φ induces a quasi-isomorphism of the E_∞-term of the spectral sequence obtained when we filter by simplicial degree. We see then that it suffices to prove that for each $\sigma = (\alpha_0, \alpha_1, \ldots, \alpha_n)$ the morphism φ induces a quasi-isomorphism

$$\mathrm{Hom}_{S_{\alpha_n}}\left(p^*_{\alpha_0\alpha_n}E^{\cdot}_{\alpha_0}, M^{\cdot}_{\alpha_n}\right) \longrightarrow \mathrm{Hom}_{S_{\alpha_n}}\left(p^*_{\alpha_0 a_n}E^{\cdot}_{\alpha_0}, N^{\cdot}_{a_n}\right)$$

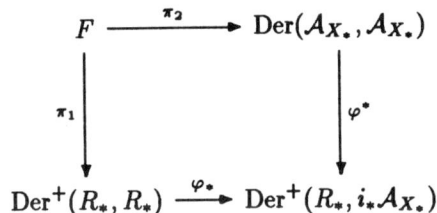

where $S_{\alpha_n} = p^*_{\alpha_0\alpha_n} R_{\alpha_0}$. We first claim that the underlying map of sheaf homomorphisms is a quasi-isomorphism. This follows immediately because $p^*_{\alpha_0\alpha_n} E^{\cdot}_{\alpha_0}$ is free over S_{α_n}. Now observe that the components of these complex of sheaf-homomorphisms are acyclic. Hence we obtain the required quasi-isomorphism from the local-to-global spectral sequence. □

Lemma 7.5. *The natural map* $d^* : \mathrm{Hom}_{R_*}(\Omega^1_{R_*}, i_*\mathcal{A}_{X_*}) \longrightarrow \mathrm{Der}(R_*, i_*\mathcal{A}_{X_*})$ *is an isomorphism.*

Proof. Let $\mathcal{V}^{\cdot}_*$ be the locally-free $\mathcal{O}_{W_*}$ module associated to the holomorphic simplicial vector bundle $V^{\cdot}_* = \bigoplus_{i\in I} p_i^* T^{1,0}(W_i) \oplus \bigoplus_{\alpha\in\mathcal{N}} p^*_\alpha(F^{\cdot}_\alpha)$. By Lemma 7.8(iv) and the adjunction formula [F], §2.1, the induced map

$$\mathrm{Hom}_{\mathcal{O}_{W_*}}\left(\mathcal{V}^{\cdot}_*, i_*\mathcal{A}^{0,\cdot}_{X_*}\right) \longrightarrow \mathrm{Der}(R_*, i_*\mathcal{A}^{0,\cdot}_{X_*})$$

is an isomorphism. The lemma follows from base-change and the isomorphism

$$\Omega^1_{R_*} \cong \mathcal{V}^{\cdot}_* \otimes_{\mathcal{O}_{W_*}} R_*.$$

□

Proposition 7.1. *The natural map* $\varphi_* : \mathrm{Der}^+(R_*, R_*) \longrightarrow \mathrm{Der}^+(R_*, i_*\mathcal{A}_{X_*})$ *is a quasi-isomorphism of complexes.*

Proof. By Lemma 7.5 it suffices to prove that

$$\varphi_* : \mathrm{Hom}_{R_*}(\Omega^1_{R_*}, R_*) \longrightarrow \mathrm{Hom}_{R_*}(\Omega^1_{R_*}, i_*\mathcal{A}_{X_*})$$

is a quasi-isomorphism. But $\Omega^1_{R_*}$ is a free R_*-module and $\varphi : R_* \longrightarrow i_*\mathcal{A}_{X_*}$ is a quasi-isomorphism. The proposition follows from Lemma 7.4. □

It is somewhat more difficult to prove that φ^* is a quasi-isomorphism. We will use the spectral sequence of Chapter 4. We consider the double complex $C^{\cdot}\left(\mathrm{sd}\,\mathcal{N}, \mathrm{Hom}_{\mathcal{O}_{X_*}}(\Omega^1_{X_*}, \mathcal{A}_{X_*})\right)$.

Lemma 7.6. $\check{H}^i\left(C^{\cdot}\left(\mathrm{sd}\,\mathcal{N}, \mathrm{Hom}_{\mathcal{O}_{X_*}}(\Omega^1_{X_*}, \mathcal{A}_{X_*})\right)\right) = 0, \quad i > 0.$

Proof. Since X is non-singular Ω^1_X is locally free. Also $\mathcal{A}$ is a complex of fine sheaves. Hence by the corollary to Theorem 4.2 we have

$$\check{H}^i\left(C^{\cdot}\left(\mathrm{sd}\,\mathcal{N}, \mathrm{Hom}_{\mathcal{O}_{X_*}}(\Omega^1_{x_*}, \mathcal{A}_{X_*})\right)\right) = H^i\left(X, \Theta \otimes_{\mathcal{O}_X} \mathcal{A}\right).$$

Since $\Theta \otimes_{\mathcal{O}_X} \mathcal{A}$ is a complex of fine sheaves the lemma follows. □

Proposition 7.2. *The natural map* $\varphi^* : \mathrm{Der}(\mathcal{A}_{X_*}, \mathcal{A}_{X_*}) \longrightarrow \mathrm{Der}(R_*, i_*\mathcal{A}_{X_*})$ *is a quasi-isomorphism of complexes.*

Proof. We have $\mathrm{Der}(\mathcal{A}_{X_*}, \mathcal{A}_{X_*}) = \mathrm{Der}(\mathcal{A}, \mathcal{A})$. By Theorem 2.2 (bis) the natural map $\iota^* : \mathrm{Der}(\mathcal{A}_{X_*}, \mathcal{A}_{X_*}) \longrightarrow \mathrm{Der}(\mathcal{O}_{X_*}, \mathcal{A}_{X_*})$ is a quasi-isomorphism of complexes. Thus it suffices to prove that $\varepsilon^* : \mathrm{Der}(\mathcal{O}_{X_*}, \mathcal{A}_{X_*}) \longrightarrow \mathrm{Der}^+(R_*, i_*\mathcal{A}_{X_*})$ is a quasi-isomorphism. We claim that it suffices to prove that the map of double complexes

$$\varepsilon^* : C^{\cdot}\left(\mathrm{sd}\,\mathcal{N}, \mathrm{Der}(\mathcal{O}_{X_*}, \mathcal{A}_{X_*})\right) \longrightarrow C^{\cdot}\left(\mathrm{sd}\,\mathcal{N}, \mathrm{Der}^+(R_*, i_*\mathcal{A}_{X_*})\right)$$

induces a quasi-isomorphism of total complexes. Indeed to prove the claim it suffices to prove that the inclusions

$$\mathrm{Der}(\mathcal{O}_{X_\bullet}, \mathcal{A}_{X_\bullet}) \longrightarrow C^\bullet(\mathrm{sd}\,\mathcal{N}, \mathrm{Der}(\mathcal{O}_{X_\bullet}, \mathcal{A}_{X_\bullet}))$$

and

$$\mathrm{Der}^+(R_*, i_*\mathcal{A}_{X_\bullet}) \longrightarrow C^\bullet\left(\mathrm{sd}\,\mathcal{N}, \mathrm{Der}^+(R_*, i_*\mathcal{A}_{X_\bullet})\right)$$

induce isomorphisms of simplicial cohomology; that is the simplicial cohomology of both double complexes in positive degree vanishes. For the first double complex this is Lemma 7.6 (combined with Lemma 2.12), for the second this is Theorem 4.3.

Thus in order to prove the proposition it suffices to prove that ε^* induces an isomorphism of the E_1-terms of the spectral sequence obtained by filtering the double complex by simplicial degree.

Thus it suffices to prove that for any pair $\alpha, \beta \in A$ with $\alpha \subset \beta$ the induced map

$$\varepsilon^* : \mathrm{Der}\left(\mathcal{O}_{X_\alpha}, (j_{\alpha\beta})_*\mathcal{A}_{X_\beta}\right) \longrightarrow \mathrm{Der}^+\left(R_\alpha, (p_{\alpha\beta}\circ i_\beta)_*\mathcal{A}_{X_\beta}\right)$$

is a quasi-isomorphism. Hence using $p_{\alpha\beta}\circ i_\beta = i_\alpha \circ j_{\alpha\beta}$ and the adjunction formula it suffices to prove that

$$\varepsilon^* : \mathrm{Hom}_{\mathcal{O}_{X_\beta}}(\Omega^1_{X_\beta}, \mathcal{A}_{X_\beta}) \longrightarrow \mathrm{Hom}_{\mathcal{O}_{X_\beta}}(j^*_{\alpha\beta}\mathcal{P}_\alpha, \mathcal{A}_{X_\beta})$$

is a quasi-isomorphism, recall $\mathcal{P}_\alpha = i_\alpha^{-1}\Omega^1_{R_\alpha} \otimes_{i^1_\alpha R_\alpha} \mathcal{O}_{X_\alpha}$.

Since $j^*_{\alpha\beta}$ is exact the complex $j^*_{\alpha\beta}\mathcal{P}_\alpha$ is a projective resolution of $\Omega^1_{X_\beta}$. But $\Omega^1_{X_\beta}$ is projective (since X_β is Stein) whence $\varepsilon : j^*_{\alpha\beta}\mathcal{P}_\alpha \longrightarrow \Omega^1_{X_\beta}$ is a homotopy equivalence. □

Corollary. *$L^\bullet_X$ and $\mathcal{L}^\bullet(X)$ are quasi-isomorphic as complexes.*

Proof. From Proposition 7.1 and 7.2 we find that L_X and $\mathrm{Der}(\mathcal{A}_{X_\bullet})$ are quasi-isomorphic as complexes. Clearly $\mathrm{Der}(\mathcal{A}_{X_\bullet}) = \mathrm{Der}(\mathcal{A}^{0,\bullet})$. But we have seen (see the discussion preceding Lemma 2.12) that $\mathrm{Der}(\mathcal{A}^{0,\bullet}) = \mathrm{Der}(\mathcal{A}^{0,\bullet}(X))$. Thus the corollary follows from Theorem 2.2. □

Unfortunately the above fiber square is not a good square, as $\varphi_* - \varphi^*$ is not onto, and more work is required to prove that $L^\bullet_X$ and $\mathcal{L}^\bullet(X)$ are quasi-isomorphic as *differential graded Lie algebras.*

To this end we let $\mathcal{C}^\infty_{W_\bullet}$ denote the sheaf of $\mathbb{C}$-algebras on W_* such that $\mathcal{C}^\infty_{W_\alpha}$ is the sheaf of complex-valued C^∞-functions on W_α. We give $\mathcal{C}^\infty_{W_\bullet}$ the obvious (inclusion of pull-back) structure maps. Then $(W_*, \mathcal{C}^\infty_{W_\bullet})$ is a simplicial ringed space to which we may apply the general theory of Chapter 4 (as well as Lemma 7.4 above). We now have a much stronger projectivity result than in the analytic case. Let $\mathcal{F}_\alpha$ be the $\mathcal{C}^\infty_{W_\alpha}$-module given by the sheaf of sections of a smooth vector bundle F_α. We claim that $p^*_\alpha\mathcal{F}_\alpha$ is a projective object in $\mathrm{Mod}(\mathcal{C}^\infty_{W_\bullet})$. Indeed by Lemma 4.1 the functor $\mathrm{Hom}_{\mathcal{C}^\infty_{W_\bullet}}(p^*_\alpha\mathcal{F}_\alpha, \cdot)$ is the composition $\mathrm{Hom}_{\mathcal{C}^\infty_{W_\alpha}}(\mathcal{F}_\alpha, \cdot) \circ \mathrm{res}_\alpha$. Clearly res_α

is exact and since the global section functor $\Gamma(W_\alpha, \cdot)$ is exact on sheaves of $\mathcal{C}^\infty_{W_\alpha}$-modules (see [Go], §3.7) the claim follows in case F_α is trivial. The general case follows by embedding F_α as a summand of a trivial bundle.

Remark. In what follows we will often use vector bundles instead of sheaves of locally free $\mathcal{O}$-modules or $\mathcal{C}^\infty$-modules. A vector bundle on W_* is defined to be a covariant functor V_* from $\mathcal{N}$ to the category of (finite dimensional) vector bundles such that V_α is over W_α and $r_{\beta\alpha} : p^*_{\alpha\beta} V_\alpha \longrightarrow V_\beta$ is a morphism of vector bundles over W_β which is injective on fibers. We do not assume $\dim V_\alpha$ is independent of α. We denote the resulting category by Vect(W$_*$). We have functors res_α and p^*_α relating Vect(W$_*$) and Vect(W$_\alpha$) corresponding to the functors on sheaves introduced in Chapter 4. A vector bundle of the form $\bigoplus_{\alpha\in\mathcal{N}} p^*_\alpha F_\alpha$ will be said to be left-induced. Note that $R^\bullet_*$ is the sheaf of holomorphic sections of $S(E^\bullet_*)$, where $E^\bullet_* = \bigoplus_{\alpha\in\mathcal{N}} p^*_\alpha F^\bullet_\alpha$ is a left-induced negatively graded holomorphic vector bundle. We note that if E_* and F_* are vector bundles on W_* with associated sheaves $\mathcal{E}_*$ and $\mathcal{F}_*$ we can express the coefficient system $\mathrm{Hom}_{\mathcal{C}^\infty_{W_*}}(\mathcal{E}_*, \mathcal{F}_*)$ of Chapter 4 in terms of the bundles E_* and F_* as follows. Define $\mathrm{Hom}(E_*, F_*)$ on a simplex $\sigma = (\alpha_0, \alpha_1, \ldots, \alpha_n)$ of sd$\,\mathcal{N}$ to be $\Gamma\left(W_{\alpha_n}, \mathrm{Hom}\left(p^*_{\alpha_0\alpha_n} E_{\alpha_0}, F_{\alpha_n}\right)\right)$. We describe the structure maps for the case of a 2-simplex $\sigma = (\alpha, \beta, \gamma)$ of sd$\,\mathcal{N}$ and leave the general case to the reader. The structure maps are all induced by bundle maps which we describe. The map $f^{(1)} : \mathrm{Hom}(p^*_{\beta\gamma} E_\beta, F_\gamma) \longrightarrow \mathrm{Hom}(p^*_{\alpha\gamma} E_\alpha, F_\gamma)$ corresponding to the inclusion $(\beta, \gamma) \subset (\alpha, \beta, \gamma)$ is given by $f^{(1)}(T) = T \circ p^*_{\beta\gamma}(r_{\beta\alpha})$. The structure map corresponding to the inclusion $(\alpha, \gamma) \subset (\alpha, \beta, \gamma)$ is the identity. The structure map corresponding to the inclusion $(\alpha, \beta) \subset (\alpha, \beta, \gamma)$ is given by $f^{(3)}(T) = s_{\gamma\beta} \circ p^*_{\beta\gamma}(T)$. Here the r (resp. s) are the structure maps of E_* (resp. F_*).

We now let $\mathcal{A}^{0,\bullet}_{W_*}$ be the differential graded $\mathcal{C}^\infty_{W_*}$-algebra such that $\mathcal{A}^{0,\bullet}_{W_\alpha}$ is the Dolbeault complex on W_α. Then $\mathcal{A}^{0,\bullet}_{W_*}$ is a free $\mathcal{C}^\infty_{W_*}$-algebra. Here we have relaxed the definition of free given in Chapter 4 to allow finitely many terms of positive degree. Indeed $\mathcal{A}^{0,\bullet}_{W_*}$ is the $\mathcal{C}^\infty_{W_*}$-algebra associated to the vector bundle $S\left(\bigoplus_{i\in I} p^*_i T^{0,1}(W_i)^*[-1]\right)$.

We define the differential bigraded algebra $\widetilde{R}^{\bullet,\bullet}_{W_*}$ by

$$\widetilde{R}^{\bullet,\bullet}_{W_*} = \mathcal{A}^{0,\bullet}_{W_*} \otimes_{\mathcal{O}_{W_*}} R^\bullet_* \,.$$

Then $\widetilde{R}^{\bullet,\bullet}_{W_*}$ is the $\mathcal{C}^\infty_{W_*}$-algebra associated to the symmetric algebra of the left-induced bundle $\bigoplus_{i\in I} p^*_i T^{0,1}(W_i)^*[-1] \oplus \bigoplus_{\alpha\in\mathcal{N}} p^*_\alpha F^\bullet_\alpha$.

We note that by Lemma 7.3 we may assume $F^\bullet_\alpha = 0$, $|\alpha| = 0$ and $\dim F^\bullet_\alpha < \infty$, all α. We observe that we have natural maps of complexes $\tilde{\varphi} : R^\bullet_* \longrightarrow \widetilde{R}^{\bullet,\bullet}_*$, $\tilde{\varepsilon} : \widetilde{R}^{\bullet,\bullet}_* \longrightarrow \mathcal{A}^{0,\bullet}_{W_*} \otimes i_* \mathcal{O}_{X_*}$ and $\tilde{\rho} : \mathcal{A}^{0,\bullet}_{W_*} \otimes i_* \mathcal{O}_{X_*} \longrightarrow i_* \mathcal{A}^{0,\bullet}_{X_*}$. In what follows we will need Malgrange's Theorem that the sheaf of smooth functions on $\mathbb{C}^n$ is flat over the sheaf of holomorphic functions (see [Ma], Chapter VI, Theorem 1.1). It follows immediately that $\mathcal{C}^{\infty}_{W_*}$ is flat over $\mathcal{O}_{W_*}$.

Lemma 7.7. *The maps $\tilde{\varphi}$, $\tilde{\varepsilon}$ and $\tilde{\rho}$ are quasi-isomorphisms.*

Proof. Since $R^{\cdot}_*$ is free over $\mathcal{O}_{W_*}$ and the inclusion $\mathcal{O}_{W_*} \longrightarrow \mathcal{A}^{0,\cdot}_{W_*}$ is a quasi-isomorphism of $\mathcal{O}_{W_*}$-modules it follows that $\tilde{\varphi}$ is a quasi-isomorphism. Since $\mathcal{A}^{0,\cdot}_{W_*}$ is free over $\mathcal{C}^{\infty}_{W_*}$ it is flat over $\mathcal{O}_{W_*}$. Since $\varepsilon : R^{\cdot}_* \longrightarrow i_*\mathcal{O}_{X_*}$ is a quasi-isomorphism, the map $\tilde{\varepsilon}$ is a quasi-isomorphism. Finally since the composition $\tilde{\rho} \circ \tilde{\varepsilon} \circ \tilde{\varphi}$ coincides with the quasi-isomorphism $\varphi = \iota \circ \varepsilon$ introduced above it follows that $\tilde{\rho}$ is a quasi-isomorphism. □

We put $\tilde{\eta} = \tilde{\rho} \circ \tilde{\varepsilon}$ whence $\varphi = \tilde{\eta} \circ \tilde{\varphi}$. Our goal is to prove that the fiber squares of complexes

$$\begin{array}{ccc} F_1 & \longrightarrow & \mathrm{Der}(\widetilde{R}^{\cdot,\cdot}, \widetilde{R}^{\cdot,\cdot}) \\ \Big\downarrow & & \Big\downarrow \tilde{\varphi}^* \\ \mathrm{Der}(R^{\cdot}, R^{\cdot}) & \xrightarrow{\tilde{\varphi}_*} & \mathrm{Der}(R^{\cdot}, \widetilde{R}^{\cdot,\cdot}) \end{array} \qquad (*)$$

and

$$\begin{array}{ccc} F_2 & \longrightarrow & \mathrm{Der}(\mathcal{A}^{0,\cdot}_{X_*}, \mathcal{A}^{0,\cdot}_{X_*}) \\ \Big\downarrow & & \Big\downarrow \tilde{\eta}^* \\ \mathrm{Der}(\widetilde{R}^{\cdot,\cdot}, \widetilde{R}^{\cdot,\cdot}) & \xrightarrow{\tilde{\eta}_*} & \mathrm{Der}(\widetilde{R}^{\cdot,\cdot}, i_*\mathcal{A}^{0,\cdot}_{X_*}) \end{array} \qquad (**)$$

are good squares. Theorem D will then follow from

$$L^{\cdot}_X = \mathrm{Der}^+(R^{\cdot}) \cong \mathrm{Der}(R^{\cdot}) \cong \mathrm{Der}(\widetilde{R}^{\cdot,\cdot}) \cong \mathrm{Der}(\mathcal{A}^{0,\cdot}_{X_*}) \cong \mathcal{L}^{\cdot}(X).$$

We now define left-induced graded vector bundles $U^{\cdot}_*$ and $V^{\cdot}_*$ on W_* by

$$\begin{aligned} U^{\cdot}_* &= \bigoplus_{i\in I} p_i^* T(W_i) \otimes \mathbb{C} \oplus \bigoplus_{i\in I} p_i^* T^{0,1}(W_i)[1] \oplus \bigoplus_{\alpha\in\mathcal{N}} p_\alpha^*(F^{\cdot}_\alpha) \\ V^{\cdot}_* &= \bigoplus_{i\in I} p_i^* T^{1,0}(W_i) \oplus \bigoplus_{\alpha\in\mathcal{N}} p_\alpha^*(F^{\cdot}_\alpha)\ . \end{aligned}$$

We then have a split exact sequence in Vect(W$_*$).

$$0 \longrightarrow V^{\cdot}_* \longrightarrow U^{\cdot}_* \longrightarrow \bigoplus_{i\in I} p_i^*(W_i)^{0,1} \oplus \bigoplus_{i\in I} p_i^* T^{0,1}(W_i)[1] \longrightarrow 0.$$

We will abbreviate the cokernel of $V^{\cdot}_* \longrightarrow U^{\cdot}_*$ to $C^{\cdot}_*$. We let $\mathcal{C}^{\cdot}_*$ denote the sheaf of smooth sections of $C^{\cdot}_*$. We note that $\mathcal{C}^{\cdot}_*$ is a free $\mathcal{C}^{\infty}_{W_*}$-module with generators concentrated on the zero skeleton of $\mathcal{N}$.

We now apply our results from Chapter 2.

Lemma 7.8. *There are isomorphisms of coefficient systems (and hence of simplicial H^0) on* sd $\mathcal{N}$

(i) $\mathrm{Hom}\left((U_*^\bullet)^*, \Lambda^\bullet T^{0,1}(W_*)^* \otimes S(E_*^\bullet)\right) \longrightarrow \mathrm{Der}(\widetilde{R}^{\bullet,\bullet}, \widetilde{R}^{\bullet,\bullet})$.
(ii) $\mathrm{Hom}\left((V_*^\bullet)^*, \Lambda^\bullet T^{0,1}(W_*)^* \otimes S(E_*^\bullet)\right) \longrightarrow \mathrm{Der}(R^\bullet, \widetilde{R}^{\bullet,\bullet})$.
(iii) $\mathrm{Hom}\left((U_*^\bullet)^*|X_*, \Lambda^\bullet T^{0,1}(X_*)^*\right) \longrightarrow \mathrm{Der}(\widetilde{R}^{\bullet,\bullet}, i_*\mathcal{A}_{X_*}^{0,\bullet})$.
(iv) $\mathrm{Hom}\left((V_*^\bullet)^*|X_*, \Lambda^\bullet T^{0,1}(X_*)^*\right) \longrightarrow \mathrm{Der}(R_*, i_*\mathcal{A}_{X_*}^{0,\bullet})$.

Proof. We need to construct compatible connections ∇_α and ∇_β on $E_\alpha^\bullet$ and $E_\beta^\bullet$ for $\alpha, \beta \in \mathcal{N}$, $\alpha \subset \beta$. We have $E_*^\bullet = \bigoplus_{\alpha \in \mathcal{N}} p_\alpha^* F_\alpha^\bullet$ whence $E_\beta^\bullet = \bigoplus_{\alpha \in \mathcal{N}} p_{\alpha\beta}^* F_\alpha^\bullet$. Clearly the vector bundle $E_*^\bullet$ can be compatibly trivialized by induction on $|\alpha|$. Hence we can give $E_*^\bullet$ the flat connection such that the trivializations are parallel. Thus the local theory of Chapter 2 applies to each pair $\alpha, \beta \in \mathcal{N}$ with $\alpha \subset \beta$. To prove equality of the above coefficient systems it suffices to prove they agree on the simplices of $\mathrm{sd}\,\mathcal{N}$ and their face maps correspond. For (i) we apply Theorem 2.3, for (ii) we apply Theorem 2.3 (bis) and for (iii) and (iv) we apply Theorem 2.3 (tertio) and Theorem 2.3 (quarto) with $\bar{E}_*^\bullet = 0$. □

Corollary. *Let $\mathcal{U}_*^\bullet$ be the locally-free $\mathcal{C}_{W_*}^\infty$–module associated to the simplicial vector bundle $U_*^\bullet$. Then the induced map*

$$\mathrm{Hom}_{\mathcal{C}_{W_*}^\infty}\left(\mathcal{U}_*^\bullet, i_*\mathcal{A}_{X_*}^{0,\bullet}\right) \longrightarrow \mathrm{Der}\left(\widetilde{R}^{\bullet,\bullet}, i_*\mathcal{A}_{X_*}^{0,\bullet}\right)$$

is an isomorphism.

Proof. The corollary follows from (iii) above by the adjunction formula [F], §2.1. □

Remark. The reader will verify that if $\mathcal{V}_*^*$ denotes the $\mathcal{O}_{W_*}$–module of holomorphic sections of V_*^*, then $\Omega_{R_*}^1 = \mathcal{V}_*^* \otimes_{\mathcal{O}_{W_*}} R_*$.

We can now prove that $(*)$ and $(**)$ are good squares.

Proposition 7.3. *The fiber square $(*)$ is a good square.*

Proof. Since $(V_*^\bullet)^*$ is a summand of $(U_*^\bullet)^*$ in $\mathrm{Vect}(\mathrm{W}_*)$, the map $\tilde{\varphi}^*$ is onto. Also $\tilde{\varphi}_*$ is a quasi-isomorphism by Lemma 7.4. It remains to check that $\ker\tilde{\varphi}^*$ is an acyclic complex. But we have by Lemma 4.1

$$\begin{aligned}
\ker\tilde{\varphi}^* &= \mathrm{Hom}\left((C_*^\bullet)^*, \Lambda^\bullet T^{0,1}(W_*)^* \otimes S(E_*^\bullet)\right) \\
&= \prod_{i \in I} \Gamma\left(W_i, \mathrm{Hom}\left(T^{0,1}(W_i)^* \otimes T^{0,1}(W_i)^*[-1], \Lambda^\bullet T^{0,1}(W_i)^*\right)\right) \\
&= \prod_{i \in I} \mathrm{Der}_{\mathcal{O}_{W_i}}\left(\mathcal{A}_{W_i}^{0,\bullet}\right).
\end{aligned}$$

The proposition follows from the corollary to Theorem 2.2 (bis). □

Proposition 7.4. *The fiber square $(**)$ is a good square.*

Proof. Since the sheaf of smooth sections of the left-induced bundle $(U^\cdot_*)^*$ is a projective $\mathcal{C}^\infty_{W_\bullet}$-module and the map $\tilde{\eta}$ is a surjective quasi-isomorphism it follows that $\tilde{\eta}_*$ is a surjective quasi-isomorphism. Since $\varphi = \tilde{\eta} \circ \tilde{\varphi}$ and φ^* is a quasi-isomorphism by Proposition 7.2 it suffices to check that the induced map $\tilde{\tilde{\varphi}}^* : \mathrm{Der}(\tilde{R}^{\cdot,\cdot}, i_*\mathcal{A}^{0,\cdot}_{X_\bullet}) \longrightarrow \mathrm{Der}(R^\cdot, i_*\mathcal{A}^{0,\cdot}_{X_\bullet})$ is a quasi-isomorphism. Since the exact sequence $\longrightarrow (C^\cdot_*)^* \longrightarrow (U^\cdot_*)^* \longrightarrow (V^\cdot_*)^* \longrightarrow 0$ is split the map $\tilde{\tilde{\varphi}}^*$ is onto and it suffices to prove that $\ker \tilde{\tilde{\varphi}}^*$ is an acyclic complex. But (noting that $W_i = X_i, i \in I$) we have

$$\begin{aligned}
\ker \tilde{\tilde{\varphi}}^* &= \mathrm{Hom}_{\mathcal{C}^\infty_{W_\bullet}}\left((C^\cdot_*)^*|X_*, \mathcal{A}^{0,\cdot}_{X_\bullet}\right)\\
&= \prod_{i\in I} \Gamma\left(W_i, \mathrm{Hom}\left(T^{0,1}(W_i)^* \oplus T^{0,1}(W_i)^*[-1], \Gamma^\cdot T^{0,1}(W_i)^*\right)\right)\\
&= \prod_{i\in I} \mathrm{Der}_{\mathcal{O}_{W_i}}(\mathcal{A}^{0,\cdot}_{W_i}).
\end{aligned}$$

The proposition follows from the corollary to Theorem 2.2 (bis). □

8. The Akahori Complexes

In [Ak3] and [Ak4], Akahori introduced complexes $(\mathcal{B}^\cdot, \bar{\partial}_b)$ — denoted $\left(\Gamma(b\Omega, E.), \bar{\partial}_b\right)$ in [Ak3] — and $(\mathcal{E}^\cdot, \bar{\partial})$ which are subcomplexes of Kuranishi's complex $(\mathcal{K}^\cdot, \bar{\partial}_b)$ on M and the Kodaira-Spencer algebra $\left(\mathcal{L}^\cdot(U), \bar{\partial}\right)$ of U respectively. We will refer to these complexes as Akahori's boundary complex and Akahori's interior complex respectively. The two complexes are related as follows (see Lemma 8.6)

$$\mathcal{E}^\cdot = \tau^{-1}\mathcal{B},$$

where $\tau : \mathcal{L}^\cdot \longrightarrow \mathcal{K}^\cdot$ is defined in Chapter 3. We define $\mathcal{B}^q$ (for $q > 0$) by

$$\mathcal{B}^q = \{\mu \in \mathcal{K}^q : \mu \lrcorner\, \theta = 0,\ \mu \lrcorner\, d\theta = 0\}.$$

Remark. We remind the reader that the notation $\alpha \lrcorner\, \beta$ was introduced in Chapter 2 and θ was defined in Chapter 3. We also recall that $\mathcal{A}^{0,q}(M) = \Gamma\left(M, \Lambda^q T^{0,1}(M)^*\right)$. We will henceforth abbreviate $\mathcal{L}^\cdot(M)$ to $\mathcal{L}^\cdot$.

We have the following result of Akahori (Theorem 2.2 of [Ak3]).

Proposition 8.1. *$(\mathcal{B}^\cdot, \bar{\partial}_b)$ is a complex.*

Using the formalism [FN1] of vector-valued forms, we can give a different proof of this proposition that illuminates the condition $\mu \lrcorner\, d\theta = 0$. The proposition will follow from the next lemma.

Lemma 8.1. *Suppose $\mu \in \mathcal{K}^q$. Then we have an equality of $(0, q+1)$-forms*

$$\bar{\partial}_b(\mu \lrcorner\, \theta) = (\bar{\partial}_b\mu) \lrcorner\, \theta + (-1)^{q-1}\mu \lrcorner\, d\theta.$$

Proof. Let $\bar{Z}_1, \ldots, \bar{Z}_q$ be smooth sections of $T^{0,1}(M)$. Then we have (see Chapter 3)

$$\begin{aligned} \bar{\partial}_b \mu\,(\bar{Z}_1, \ldots, \bar{Z}_{q+1}) &= \sum_{i=1}^{q+1} (-1)^{i-1} P\left([\bar{Z}_i, \mu(\bar{Z}_1, \ldots, \hat{\bar{Z}}_i, \ldots, \bar{Z}_{q+1})]\right) + \\ & \sum_{1 \le i < j \le q+1} (-1)^{i+j} \mu\left([\bar{Z}_i, \bar{Z}_j], \ldots, \hat{\bar{Z}}_i, \ldots, \hat{\bar{Z}}_j, \ldots, \hat{\bar{Z}}_{q+1}\right). \end{aligned}$$

Hence

$$\begin{aligned} \bar{\partial}_b \mu \lrcorner\, \theta\,(\bar{Z}_1, \ldots, \bar{Z}_{q+1}) = & \sum_{i=1}^{q+1} (-1)^{i-1} \theta\left([\bar{Z}_i, \mu(\bar{Z}_1, \ldots, \hat{\bar{Z}}_i, \ldots, \bar{Z}_{q+1})]\right) + \\ & \sum_{1 \le i < j \le q+1} (-1)^{i+j} \theta\left(\mu([\bar{Z}_i, \bar{Z}_j], \ldots, \hat{\bar{Z}}_i, \ldots, \hat{\bar{Z}}_j, \ldots, \bar{Z}_{q+1})\right). \end{aligned}$$

Also we have

$$\begin{aligned} \mu \lrcorner\, d\theta\,(\bar{Z}_1, \ldots, \bar{Z}_{q+1}) &= \sum_{i=1}^{q+1} (-1)^{q+i-1} d\theta\left(\mu(\bar{Z}_1, \ldots, \hat{\bar{Z}}_i, \ldots, \bar{Z}_{q+1}), \bar{Z}_i\right) \\ &= (-1)^q \sum_{i=1}^{q+1} (-1)^i\, \bar{Z}_i\, \theta\left(\mu(\bar{Z}_1, \ldots, \hat{\bar{Z}}_i, \ldots, \bar{Z}_{q+1})\right) + \\ & \quad (-1)^q \sum_{i=1}^{q+1} (-1)^i \theta\left([\mu(\bar{Z}_1, \ldots, \hat{\bar{Z}}_i, \ldots, \bar{Z}_{q+1}), \bar{Z}_i]\right). \end{aligned}$$

Thus

$$\begin{aligned} & \left(\bar{\partial}_b \mu \lrcorner\, \theta + (-1)^{q-1} \mu \lrcorner\, d\theta\right)(\bar{Z}_1, \ldots, \bar{Z}_{q+1}) \\ &= \sum_{i=1}^{q+1} (-1)^{i-1}\, \bar{Z}_i\, \theta\left(\mu(\bar{Z}_1, \ldots, \hat{\bar{Z}}_i, \ldots, \bar{Z}_{q+1})\right) + \\ & \quad \sum_{1 \le i < j \le q+1} (-1)^{i+j} \theta\left(\mu([\bar{Z}_i, \bar{Z}_j], \ldots, \hat{\bar{Z}}_i, \ldots, \hat{\bar{Z}}_j, \ldots, \bar{Z}_{q+1})\right) \\ &= \bar{\partial}_b(\mu \lrcorner\, \theta)\,(\bar{Z}_1, \ldots, \bar{Z}_{q+1}). \end{aligned}$$

□

Corollary. *If $\mu \lrcorner\, \theta = 0$ then $(\bar{\partial}_b \mu) \lrcorner\, \theta = (-1)^q \mu \lrcorner\, d\theta$.* □

Corollary. *$(\mathcal{B}^{\bullet}, \bar{\partial}_b)$ is a complex.*

Proof. Let $\mu \in \mathcal{B}^q$. We must prove $\bar{\partial}_b \mu \lrcorner\, \theta = 0$ and $\bar{\partial}_b \mu \lrcorner\, d\theta = 0$. But $\mu \in \mathcal{B}^q$ implies $\mu \lrcorner\, \theta = 0$ and $\mu \lrcorner\, d\theta = 0$ whence $\partial_b \mu \lrcorner\, \theta = 0$ by the above lemma. Also by the above lemma since $\bar{\partial}_b \mu \lrcorner\, \theta = 0$ we have

$$\bar{\partial}_b \mu \lrcorner\, d\theta = (-1)^{q+1} \bar{\partial}_b^2 \mu \lrcorner\, \theta = 0.$$

□

Corollary. *Suppose* $\mu \in \mathcal{A}^{0,q}\left(M, T^{1,0}(M)\right)$, $q > 0$, *and* $\bar{\partial}_b \mu \in \mathcal{A}^{0,q+1}\left(M, T^{1,0}(M)\right)$. *Then* $\mu \in \mathcal{B}^q$.

In [Ak3], Akahori defined $\mathcal{B}^0 = \{0\}$. For our purpose it is more useful to redefine $\mathcal{B}^0$ as follows

$$\mathcal{B}^0 = \left\{\lambda \in K^0 : (\bar{\partial}_b \lambda) \lrcorner\, \theta = 0\right\}.$$

Lemma 8.2. $\mathcal{B}^0$ *is the inverse image of* $\mathcal{B}^1 \subset K^1$ *under* $\bar{\partial}_b : \mathcal{K}^0 \longrightarrow \mathcal{K}^1$.

Proof. We have to check that $(\bar{\partial}_b \lambda) \lrcorner\, d\theta = 0$. However, this follows as above by applying $\bar{\partial}_b$ to the identity $\bar{\partial}_b \lambda \lrcorner\, \theta = 0$. □

We digress to further justify our choice of $\mathcal{B}^0$. We learned the next lemma from Jack Lee.

Lemma 8.3. $\lambda \in \mathcal{B}^0$ *if and only if there exists a smooth complex-valued function* f *such that*

$$\lambda = fT - Z_f$$

where Z_f *is the unique section of* $T^{1,0}(M)$ *satisfying*

$$Z_f \lrcorner\, d\theta = \bar{\partial}_b f.$$

Proof. We first assume λ satisfies $\bar{\partial}_b \lambda \lrcorner\, \theta = 0$. We may write $\lambda = fT - Z$ with Z a section of $T^{1,0}(M)$. Clearly $f = \lambda \lrcorner\, \theta$. We have then by Lemma 8.1

$$\bar{\partial}_b f = \bar{\partial}_b \lambda \lrcorner\, \theta - \lambda \lrcorner\, d\theta = Z \lrcorner\, d\theta.$$

Conversely assume $\lambda = fT - Z_f$ as above. Then differentiating the identity $\bar{\partial}_b \lambda \lrcorner\, \theta = f$ we obtain

$$\bar{\partial}_b f = \bar{\partial}_b \lambda \lrcorner\, \theta - \lambda \lrcorner\, d\theta = \bar{\partial}_b \lambda \lrcorner\, \theta + Z_f \lrcorner\, d\theta.$$

□

Remark. By Lemma 8.3 we can replace $\mathcal{B}^0$ by the space $C^\infty(M, \mathbb{C})$ of smooth complex-valued function on M. We note that the induced differential is a *second-order* differential operator. Thus the elements of $\mathcal{B}^q$, $q \geq 0$, can be interpreted as sections of vector bundles.

We next review a result of Akahori (essentially Theorem 2.4 of [Ak3]) which will be of great importance to use.

Proposition 8.2. *The inclusion* $i : (\mathcal{B}^\cdot, \bar{\partial}_b) \longrightarrow (\mathcal{K}^\cdot, \bar{\partial}_b)$ *is a quasi-isomorphism of complexes.*

Proof. The Proposition follows immediately from Theorem 2.2 of [Ak2]. Indeed in the proof of that theorem Akahori shows that given $\mu \in \mathcal{K}^q$, $q \geq 1$, there exists $\theta_\mu \in \mathcal{A}^{0,q-1}\left(M, T^{1,0}(M)\right)$ such that

$$\mu - \bar{\partial}_b \theta_\mu \in \mathcal{A}^{0,q}\left(M, T^{1,0}(M)\right).$$

It is immediate from this that the above inclusion induces isomorphisms on cohomology of degrees two or larger and a surjection on cohomology of degree 1. But if $\mu \in \mathcal{B}^1$ satisfies $\mu = \bar{\partial}_b \eta$ with $\eta \in \mathcal{K}^0$ then $\bar{\partial}_b \eta \in \mathcal{B}^0$ so by Lemma 8.2 we have $\eta \in \mathcal{B}^0$. Consequently i induces an isomorphism on first cohomology as well. By definition every zero cycle in $\mathcal{K}$ is in $\mathcal{B}^0$ and the proposition follows. □

We next review the definition and properties of Akahori's interior complex $(\mathcal{E}^{\cdot}, \bar{\partial})$, [Ak4]. Let ρ be the defining function of Chapter 3. We define $\mathcal{E}^q \subset \mathcal{A}^{0,q}\left(U, T^{1,0}(U)\right)$, $q > 0$, by

$$\mathcal{E}^q = \left\{\mu \in \mathcal{A}^{0,q}(U, T^{1,0}(U)) : j^*(\mu \lrcorner\, \partial\rho) = j^*(\mu \lrcorner\, \partial\bar{\partial}\rho) = 0\right\}.$$

We call that j^* was defined in Chapter 3.

We define $\mathcal{E}^0$ (again we change Akahori's definition) by

$$\mathcal{E}^0 = \left\{\lambda \in \mathcal{A}^0(U, T^{1,0}(U)) : j^*(\bar{\partial}\lambda \lrcorner\, \partial\rho) = 0\right\}.$$

We now give the analogue of Lemma 8.2.

Lemma 8.4. *The vector space $\mathcal{E}^0$ is the inverse image of $\mathcal{E}^1 \subset \mathcal{L}^1$ under $\bar{\partial} : \mathcal{L}^0 \longrightarrow \mathcal{L}^1$.*

Proof. We have to check that $\lambda \in \mathcal{E}^0$ implies $j^*(\bar{\partial}\lambda \lrcorner\, \partial\bar{\partial}\rho) = 0$. But applying $\bar{\partial}_b$ to the formula $j^*(\bar{\partial}\lambda \lrcorner\, \partial\rho) = 0$ we obtain by definition

$$0 = \bar{\partial}_b j^*(\bar{\partial}\lambda \lrcorner\, \partial\rho) = j^*\left(\bar{\partial}(\bar{\partial}\lambda \lrcorner\, \partial\rho)\right) = j^*(\bar{\partial}\lambda \lrcorner\, \bar{\partial}\partial\rho).$$

□

The next proposition is Proposition 3.1 of [Ak4].

Proposition 8.3. *$(\mathcal{E}^{\cdot}, \bar{\partial})$ is a sub-complex of $(\mathcal{L}^{\cdot}, \bar{\partial})$.*

Proof. Let $\mu \in \mathcal{E}^q$. We must show $j^*(\bar{\partial}\mu \lrcorner\, \partial\rho) = 0$ and $j^*(\bar{\partial}\mu \lrcorner\, \partial\bar{\partial}\rho) = 0$. But $j^*(\bar{\mu} \lrcorner\, \partial\rho) = 0$. Applying $\bar{\partial}_b$ to this identity, we obtain

$$\begin{aligned} 0 &= \bar{\partial}_b j^*(\mu \lrcorner\, \partial\rho) = j^*\left(\bar{\partial}(\mu \lrcorner\, \partial\rho)\right) \\ &= j^*\left(\bar{\partial}\mu \lrcorner\, \partial\rho + (-1)^{q-1}\mu \lrcorner\, \bar{\partial}\partial\rho\right). \end{aligned}$$

But $j^*(\mu \lrcorner\, \bar{\partial}\partial\rho) = 0$ since $\mu \in \mathcal{E}^q$. Now apply $\bar{\partial}_b$ to the identity $j^*(\bar{\partial}\mu \lrcorner\, \partial\rho) = 0$ to deduce $j^*(\bar{\partial}\mu \lrcorner\, \bar{\partial}\partial\rho) = 0$. □

Lemma 8.5. *Suppose $\mu \in \mathcal{L}^q$, $q > 0$, then μ satisfies $j^*(\mu \lrcorner\, \partial\rho) = 0$ and $j^*(\bar{\partial}\mu \lrcorner\, \partial\rho) = 0$ if and only if $\mu \in \mathcal{E}^q$.*

Proof. It suffices to prove that if $j^*(\mu \lrcorner\, \partial\rho) = 0$ then $j^*(\mu \lrcorner\, \partial\bar{\partial}\rho) = 0$ if and only if $j^*(\bar{\partial}\mu \lrcorner\, \bar{\partial}\rho) = 0$. We apply $\bar{\partial}_b$ to the identity $j^*(\mu \lrcorner\, \partial\rho) = 0$ to obtain $j^*(\bar{\partial}\mu \lrcorner\, \partial\rho) = (-1)^q j^*(\mu \lrcorner\, \bar{\partial}\partial\rho)$. □

Lemma 8.6. *Let $\mu \in \mathcal{L}^q$. Then*

$$j^*(\mu \lrcorner\, \partial\rho) = \tfrac{i}{2}(\tau\mu) \lrcorner\, \theta.$$

Proof. It suffices to prove the above formula in the case μ is decomposable, that is, $\mu = \omega \otimes W$ with $\omega \in \mathcal{A}^{0,q}(U)$ and $W \in \Gamma\left(U, T^{1,0}(U)\right)$. In this case we have by Lemma 2.2

$$\begin{aligned} j^*(\mu \lrcorner\, \partial\rho) &= j^*(\omega \wedge \iota_W \partial\rho) = \partial\rho(W) j^*\omega \\ (\tau\mu) \lrcorner\, \theta &= j^*\omega \wedge \iota_{\iota W}\theta = \theta(\tau W) j^*\omega. \end{aligned}$$

The lemma follows by considering the case $W = \xi$, $W \in T^{1,0}(M)$ and noting $\tau\xi = -2iT$. □

We can now prove that $(\mathcal{B}^\cdot, \bar\partial_b)$ is the complex on M induced by $(\mathcal{E}, \bar\partial)$ via τ.

Lemma 8.7. *Let $\mu \in \mathcal{L}^q$. Then $\mu \in \mathcal{E}^q$ if and only if $\tau\mu \in \mathcal{B}^q$.*

Proof. Suppose $q > 0$. Then $\mu \in \mathcal{E}^q$ if and only if $j^*(\mu \lrcorner\, \partial\rho) = 0$ and $j^*(\bar\partial\mu \lrcorner\, \partial\rho) = 0$. By the previous lemma the latter conditions hold if and only if $\tau\mu \lrcorner\, \theta = 0$ and $\tau(\bar\partial\mu) \lrcorner\, \theta = 0$. By Lemma 3.3 we have

$$\tau(\bar\partial\mu) = \bar\partial_b(\tau\mu).$$

Hence $\mu \in \mathcal{E}^q$ if and only if $\tau\mu \lrcorner\, \theta = 0$ and $\bar\partial_b(\tau\mu) \lrcorner\, \theta = 0$. But by the corollary to Lemma 8.1 we have $\bar\partial_b(\tau\mu) \lrcorner\, \theta = (-1)^q(\tau\mu) \lrcorner\, d\theta$. We leave the case $q = 0$ to the reader. □

Now we can prove the analogue of Proposition 8.2 (this is Theorem 3.4 of [Ak4]).

Proposition 8.4. *The inclusion $(\mathcal{E}^\cdot, \bar\partial) \longrightarrow (\mathcal{L}^\cdot, \bar\partial)$ is a quasi-isomorphism of complexes.*

Proof. Let $\mu \in \mathcal{L}^q$, $q \geq 1$. We will show there exists $\theta_\mu \in \mathcal{L}^{q-1}$ such that $j^*\left((\mu - \bar\partial\theta_\mu) \lrcorner\, \partial\rho\right) = 0$. Indeed let $\nu = \tau\mu$ and θ_ν be the element of $\mathcal{K}^{q-1}$ described in Proposition 8.2 whence $(\nu - \bar\partial_b\theta_\nu) \lrcorner\, \theta = 0$. Let θ_μ be any element of $\mathcal{L}^{q-1}$ with $\tau\theta_\mu = \theta_\nu$. Then $\tau(\mu - \bar\partial\theta_\mu) \lrcorner\, \theta = (\nu - \bar\partial_b\theta_\nu) \lrcorner\, \theta = 0$. Hence by Lemma 8.5 we have $j^*\left((\mu - \bar\partial\theta_\mu) \lrcorner\, \partial\rho\right) = 0$. We now prove that the above inclusion is surjective on cohomology of degree greater than 1. Indeed suppose $\mu \in \mathcal{L}^q$, $q \geq 1$, with $\bar\partial\mu = 0$. Then $\eta = \mu - \bar\partial\theta_\mu$ satisfies $j^*(\eta \lrcorner\, \partial\rho) = 0$ and $\bar\partial\eta = 0$ hence $\eta \in \mathcal{E}^q$ by Lemma 8.5. We next prove that the above inclusion is injective on cohomology of degree greater than one. Indeed suppose $\mu \in \mathcal{E}^q$, $q \geq 2$ and suppose $\eta \in \mathcal{L}^{q-1}$ satisfies $\bar\partial\eta = \mu$. Then $\bar\partial(\eta - \bar\partial\theta_\eta) = \mu$. But $j^*\left((\eta - \bar\partial\theta_\eta) \lrcorner\, \partial\rho\right) = 0$ and $\bar\partial\eta \in \mathcal{E}^q$ whence $\eta - \bar\partial\theta_\eta \in \mathcal{E}^{q-1}$. The rest of the proposition is obvious by Lemma 8.4. □

Proposition 8.5. *The restriction map $\tau : \mathcal{E} \longrightarrow \mathcal{B}$ induces isomorphisms on cohomology groups of degree i, $1 \leq i \leq n-2$, where $\dim U = n$.*

Proof. We have a commutative diagram of complexes

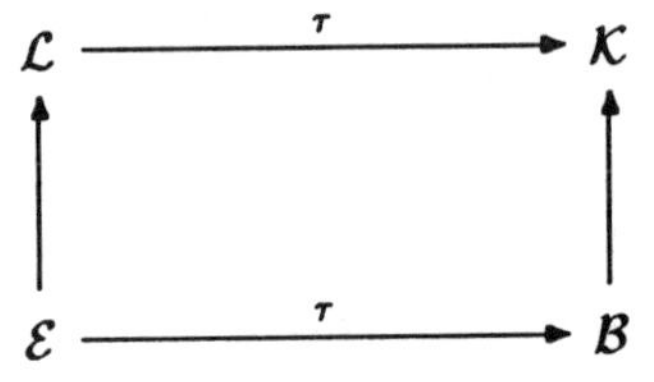

We have seen that the two vertical maps are quasi-isomorphisms and the top horizontal arrow induces an isomorphism of cohomology in the given degrees by Theorem 3.1. □

9. A Controlling Differential Graded Lie Algebra for Kuranishi's CR-deformation Theory

In the previous section we reviewed the definitions and properties of Akahori's boundary complex $(\mathcal{B}^{\cdot}, \bar{\partial}_b)$ and Akahori's interior complex $(\mathcal{E}^{\cdot}, \bar{\partial})$ and proved the critical result that $\tau : \mathcal{E}^{\cdot} \longrightarrow \mathcal{B}^{\cdot}$ induced an isomorphism of cohomology of degree i, $1 \leq i \leq n-2$. We will show that $\mathcal{E}^{+} = \bigoplus_{q=1}^{n} \mathcal{E}^{q}$ is closed under the Nijenhuis bracket. Moreover if we define $\mathcal{B}^{+} = \bigoplus_{q=1}^{n-1} \mathcal{B}^{q}$ and let $\tau^{+} : \mathcal{E}^{+} \longrightarrow \mathcal{B}^{+}$ be the map induced by τ we show that $\mathcal{J}^{+} = \ker \tau^{+}$ is an ideal in $\mathcal{E}^{+}$. Thus $\mathcal{B}^{+}$ has the structure of a differential graded Lie algebra such that $\tau^{+} : \mathcal{E}^{+} \longrightarrow \mathcal{B}^{+}$ is a (surjective) homomorphism of differential graded Lie algebras. We then "truncate" $\mathcal{B}^{+}$ to get a controlling differential graded Lie algebra $\bar{\mathcal{K}}^{\cdot}(M)$ for Kuranishi's CR-deformation theory. Here we use Miyajima's equations [Mi1] for the versal family.

In the following $Z_1, Z_2, \ldots, Z_{p+q}$ are smooth sections of $T^{1,0}(U)$ which along M take values in $T^{1,0}(M)$. If α is a $(p+q)$–linear (over $\mathcal{A}^{0,0}(U)$) function on $T^{0,1}(U)$ we define $\operatorname{Alt}\alpha$ by $\operatorname{Alt}\alpha\left(\bar{Z}_1, \ldots, \bar{Z}_{p+q}\right) = \sum_{\sigma \in G} \varepsilon(\sigma)\alpha\left(\bar{Z}_{\sigma(1)}, \ldots, \bar{Z}_{\sigma(p+q)}\right)$. Here G is the group of permutations of $\{1, 2, \ldots, p+q\}$ and $\varepsilon(\sigma)$ is the sign of the permutation σ.

Lemma 9.1. *Suppose $\mu \in \mathcal{E}^{p}$ and $\eta \in \mathcal{E}^{q}$ with $p > 0$ and $q > 0$. Then*

$$j^{*}\left([\mu, \eta] \lrcorner\, \partial\rho\right) = 0.$$

Proof. We have along M by the Frölicher-Nijenhuis formulas of Chapter 2

$$\begin{aligned}
&[\mu, \eta] \lrcorner\, \partial\rho\left(\bar{Z}_1, \ldots, \bar{Z}_{p+q}\right) \\
&\quad = \partial\rho\left([\mu, \eta](\bar{Z}_1, \ldots, \bar{Z}_{p+q})\right) \\
&\quad = \frac{1}{p!q!}\operatorname{Alt}\partial\rho\left([\mu(\bar{Z}_1, \ldots, \bar{Z}_p), \eta(\bar{Z}_{p+1}, \ldots, \bar{Z}_{p+q})]\right) \\
&\qquad + \frac{(-1)^{pq+q+1}}{(p-1)!q!}\operatorname{Alt}\partial\rho\Big(\mu\left([\bar{Z}_1, \eta(\bar{Z}_2, \ldots, \bar{Z}_{q+1})], \bar{Z}_{q+2}, \ldots, \bar{Z}_{p+q}\right)\Big) \\
&\qquad + \frac{(-1)^{p}}{p!(q-1)!}\operatorname{Alt}\partial\rho\Big(\eta\left([\bar{Z}_1, \mu(\bar{Z}_2, \ldots, \bar{Z}_{p+1})], \bar{Z}_{p+2}, \ldots, \bar{Z}_{p+q}\right)\Big).
\end{aligned}$$

The first term is zero because $\mu|\Lambda^{p}T^{0,1}(M)^{*}$ and $\eta|\Lambda^{q}T^{0,1}(M)^{*}$ take values in $T^{1,0}(M)$ and $T^{1,0}(M)$ is involutive. We now prove that the second and third terms are zero. By symmetry it suffices to prove that the second term is zero.

We first note the congruence along M, for any permutation σ,

$$[\bar{Z}_{\sigma(1)}, \eta(\bar{Z}_{\sigma(2)}, \ldots, \bar{Z}_{\sigma(q+1)})] \equiv$$
$$\bar{\partial}\rho\left([\bar{Z}_{\sigma(1)}, \eta(\bar{Z}_{\sigma(2)}, \ldots, \bar{Z}_{\sigma(q+1)})]\right)\bar{\xi} \bmod \left(T^{0,1}(M) + T^{1,0}(U)|M\right).$$

Since μ has type $(0,p)$ and $\mu|\Lambda^p T^{0,1}(M)$ takes values in $T^{1,0}(M)$ we have

$$\mu\left([\bar{Z}_{\sigma(1)}, \eta(\bar{Z}_{\sigma(2)}, \ldots, \bar{Z}_{\sigma(q+1)})], \bar{Z}_{\sigma(q+2)}, \ldots, \bar{Z}_{\sigma(p+q)}\right) \equiv$$
$$\bar{\partial}\rho\left([\bar{Z}_{\sigma(1)}, \eta(\bar{Z}_{\sigma(2)}, \ldots, \bar{Z}_{\sigma(q+1)})]\right)\mu\left(\bar{\xi}, \bar{Z}_{\sigma(q+2)}, \ldots, \bar{Z}_{\sigma(p+q)}\right) \bmod T^{1,0}(M)$$

and the second term may be rewritten

$$\frac{(-1)^{pq+q+1}}{(p-1)!q!} \sum_{\sigma \in G/H} \left\{ \sum_{\tau \in H'} \bar{\partial}\rho\left([\bar{Z}_{\sigma\tau(1)}, \eta(\bar{Z}_{\sigma\tau(2)}, \ldots, \bar{Z}_{\sigma\tau(q+1)})]\right) \times \right.$$
$$\left. \sum_{\omega \in H''} \partial\rho\left(\mu(\bar{\xi}, \bar{Z}_{\sigma\omega(q+2)}, \ldots, \bar{Z}_{\sigma\omega(p+q)})\right)\right\}.$$

Hence G is the group of permutations of $\{1, 2, \ldots, p+q\}$ and H' (resp. H'') is the subgroup that acts trivially on $q+2, \ldots, p+q$ (resp. $1, 2, \ldots, q+1$) and $H = H' \times H''$.

We will show that for each σ (and ω) the corresponding sum over H' is zero. We claim

$$\sum_{\tau \in H'} \bar{\partial}\rho\left([\bar{Z}_{\sigma\tau(1)}, \eta(\bar{Z}_{\sigma\tau(2)}, \ldots, \bar{Z}_{\sigma\tau(q+1)})]\right) = q!\eta \lrcorner\, \partial\bar{\partial}\rho\left(\bar{Z}_{\sigma(2)}, \ldots, \bar{Z}_{\sigma(q+1)}, \bar{Z}_{\sigma(1)}\right).$$

Indeed put $Y_i = \bar{Z}_{\sigma(i)}$, $1 \leq i \leq q+1$. Then $Y_{\tau(i)} = \bar{Z}_{\sigma(\tau(i))}$ so

$$\begin{aligned}
\sum_{\tau \in H'} \bar{\partial}\rho\left([\bar{Z}_{\sigma\tau(1)}, \eta(\bar{Z}_{\sigma\tau(2)}, \ldots, \bar{Z}_{\sigma\tau(q+1)})]\right) &= \operatorname{Alt} \bar{\partial}\rho\left([\mathrm{Y}_1, \eta(\mathrm{Y}_2, \ldots, \mathrm{Y}_{q+1})]\right) \\
&= -\operatorname{Alt} \partial\bar{\partial}\rho\left(\mathrm{Y}_1, \eta(\mathrm{Y}_2, \ldots, \mathrm{Y}_{q+1})\right) \\
&= q!\eta \lrcorner\, \partial\bar{\partial}\rho\left(Y_2, \ldots, Y_{q+1}, Y_1\right).
\end{aligned}$$

The claim follows.

The last expression is zero because $\eta \in \mathcal{E}^q$. □

Lemma 9.2. *Suppose $\mu \in \mathcal{E}^p$, $\eta \in \mathcal{E}^q$ with $p > 0$ and $q > 0$. Then $[\mu, \eta] \in \mathcal{E}^{p+q}$.*

Proof. We have seen that $j^*\left([\mu, \eta] \lrcorner\, \partial\rho\right) = 0$. It remains to check that $j^*\left(\bar{\partial}[\mu, \eta] \lrcorner\, \partial\rho\right) = 0$. But $\mathcal{E}$ is a complex so $\bar{\partial}\mu$ and $\bar{\partial}\eta$ are in $\mathcal{E}$ and the result follows from Lemma 9.1 since $\bar{\partial}[\mu, \eta] = [\bar{\partial}\mu, \eta] + (-1)^p[\mu, \bar{\partial}\eta]$. □

We have obtained the following theorem.

Theorem 9.1. *$(\mathcal{E}^+, \bar{\partial})$ is a differential graded Lie algebra.*

We now give the positive elements of Akahori's boundary complex $(\mathcal{B}^\cdot, \bar{\partial}_b)$ the structure of a differential graded Lie algebra such that τ preserves brackets.

We define $\mathcal{J}^\cdot \subset \mathcal{E}^\cdot$ by $\mathcal{J}^p = \operatorname{Ker}(\tau : \mathcal{E}^p \longrightarrow \mathcal{B}^p)$ and $\mathcal{J}^+ = \bigoplus_{p \geq 1} \mathcal{J}^p$.

Lemma 9.3. *$\mathcal{J}^+$ is an ideal in $\mathcal{E}^+$.*

Proof. Suppose $\mu \in \mathcal{J}^p$ and $\eta \in \mathcal{E}^q$ and let $Z_1, Z_2, \ldots, Z_{p+q}$ as above. Then along M we have

$$\begin{aligned}[\mu, \eta]\,(\bar{Z}_1, \ldots, \bar{Z}_{p+q}) = \;& \tfrac{1}{p!q!}\mathrm{Alt}\,\left[\mu(\bar{Z}_1, \ldots, \bar{Z}_p), \eta(\bar{Z}_{p+1}, \ldots, \bar{Z}_{p+q})\right] + \\ & \tfrac{(-1)^{pq+q+1}}{(p-1)!q!}\mathrm{Alt}\,\mu\left(\left[\bar{Z}_1, \eta(\bar{Z}_2, \ldots, \bar{Z}_{q+1})\right], \bar{Z}_{q+2}, \ldots, \bar{Z}_{p+q}\right) + \\ & \tfrac{(-1)^p}{p!(q-1)!}\mathrm{Alt}\,\eta\left(\left[\bar{Z}_1, \mu(\bar{Z}_2, \ldots, \bar{Z}_{p+1})\right], \bar{Z}_{p+2}, \ldots, \bar{Z}_{p+q}\right).\end{aligned}$$

The first and third terms are clearly zero since they involve tangential derivatives of the restriction of μ to M which is identically zero. It remains to show that the second term is zero.

As before we have the congruence for any permutation σ

$$\begin{aligned}&\left[\bar{Z}_{\sigma(1)}, \eta(\bar{Z}_{\sigma(2)}, \ldots, \bar{Z}_{\sigma(q+1)})\right] \equiv \\ &\qquad \bar{\partial}\rho\left(\left[\bar{Z}_{\sigma(1)}, \eta(\bar{Z}_{\sigma(2)}, \ldots, \bar{Z}_{\sigma(q+1)})\right]\right)\bar{\xi} \bmod \left(T^{0,1}(M) + T^{1,0}(U)|M\right).\end{aligned}$$

Hence, since $\mu|\Lambda^p T^{0,1}(M) = 0$ and μ is of type $(0,p)$ we have (using the notation and argument of Lemma 9.1)

$$\begin{aligned}&\mathrm{Alt}\,\mu\left(\left[\bar{Z}_1, \eta(\bar{Z}_2, \ldots, \bar{Z}_{q+1})\right], \bar{Z}_{q+2}, \ldots, \bar{Z}_{p+q}\right) \\ &= \mathrm{Alt}\left\{\bar{\partial}\rho\left(\left[\bar{Z}_1, \eta(\bar{Z}_2, \ldots, \bar{Z}_{q+1})\right]\right)\mu\left(\bar{\xi}, \bar{Z}_{q+2}, \ldots, \bar{Z}_{p+q}\right)\right\} \\ &= -\sum_{\sigma \in G/H}\left\{\sum_{\tau \in H'} \partial\bar{\partial}\rho\left(\bar{Z}_{\sigma\tau(1)}, \eta(\bar{Z}_{\sigma\tau(2)}, \ldots, \bar{Z}_{\sigma\tau(q+1)})\right)\cdot \right. \\ &\qquad \left. \sum_{\omega \in H''} \mu(\bar{\xi}, \bar{Z}_{\sigma\omega(q+2)}, \ldots, \bar{Z}_{\sigma\omega(p+q)})\right\}.\end{aligned}$$

The last expression is zero because the sum over H' is (up to sign)

$$q!\left(\eta \lrcorner\, \partial\bar{\partial}\rho(\bar{Z}_{\sigma(1)}, \ldots, \bar{Z}_{\sigma(q+1)})\right).$$

□

Remark. We obtain a graded bracket on $\mathcal{B}^+$ as follows. Let $\mu, \eta \in \mathcal{B}^+$. Choose $\tilde{\mu}, \tilde{\eta} \in \mathcal{E}^+$ with $\tau\tilde{\mu} = \mu, \tau\tilde{\eta} = \eta$. Then define $[\mu, \eta]$ by

$$[\mu, \eta] = \tau[\tilde{\mu}, \tilde{\eta}].$$

Clearly $(\mathcal{B}^+, [\ ,\ ], \bar{\partial}_b)$ is a differential graded Lie algebra with $\tau : \mathcal{E}^+ \longrightarrow \mathcal{B}^+$ a homomorphism of differential graded Lie algebras.

We now choose a complement $\bar{\mathcal{C}}^1$ to $\bar{\partial}_b\mathcal{B}^0$ in $\mathcal{B}^1$. We define a differential graded Lie algebra $\bar{\mathcal{K}} \subset \mathcal{B}^+$ by

$$\bar{\mathcal{K}} = \bar{\mathcal{C}}^1 \oplus \bigoplus_{i \geq 2} \mathcal{B}^i.$$

Remark. The elements of $\bar{\mathcal{K}}^1$ correspond to changes of the CR-structure on M keeping fixed the underlying contact structure.

The following lemma is an immediate consequence of the surjectivity of $\tau : \mathcal{E} \longrightarrow \mathcal{B}$, see [M3], Lemma 6.5.

Lemma 9.4. *There exists a complement* C^1 *to* $\bar{\partial}\mathcal{E}^0$ *in* $\mathcal{E}^1$ *such that*

$$\tau C^1 = \bar{C}^1. \qquad \square$$

We then define $\mathcal{L}_{\tan} \subset \mathcal{E}$ by

$$\mathcal{L}_{\tan} = C^1 \oplus \bigoplus_{i \geq 2} \mathcal{E}^i.$$

The next theorem is then immediate from the above and the fact that the above truncation construction does not change the cohomology groups of positive degree.

Theorem 9.2. *The restriction map* $\tau : \mathcal{L}_{tan} \longrightarrow \bar{\mathcal{K}}$ *induces an isomorphism on first cohomology if* $\dim V \geq 3$ *and an isomorphism on first and second cohomology if* $\dim V \geq 4$. $\square$

From the diagram of differential graded Lie algebras

$$\mathcal{L} \longleftarrow \mathcal{L}_{\tan} \longleftarrow \bar{\mathcal{K}}$$

and Proposition 8.4 and the above Theorem we obtain the first of the two main results of this section.

THEOREM E. *If* $\dim V \geq 4$ *there is a one-quasi-isomorphism from* $\mathcal{L}$ *to* $\bar{\mathcal{K}}$. $\square$

We can now show that $\bar{\mathcal{K}}$ is a controlling differential graded Lie algebra for Kuranishi's CR-deformation theory.

We first check that Akahori's integrability condition for elements in $\bar{\mathcal{K}}^1$ agrees with he equations studied in Chapter 1.

Lemma 9.5. *Suppose* $\varphi \in \bar{\mathcal{K}}^1$. *Then*

$$S(\varphi) = \bar{\partial}_b\varphi + \tfrac{1}{2}[\varphi, \varphi].$$

Proof. Let $\tilde{\varphi} \in \mathcal{E}^1$ be an extension of φ to a neighbourhood of M and Z and W be smooth sections of $T^{1,0}(M)$. Then by definition of [,] we have

$$\begin{aligned}
[\varphi,\varphi](\bar{Z},\bar{W}) &= [\tilde{\varphi},\tilde{\varphi}](\bar{Z},\bar{W}) \\
&= 2\left[\varphi(\bar{Z}),\varphi(\bar{W})\right] - 2\tilde{\varphi}\left([\bar{Z},\varphi(\bar{W})]\right) + 2\tilde{\varphi}\left([\bar{W},\varphi(\bar{Z})]\right) \\
&= 2\left[\varphi(\bar{Z}),\varphi(\bar{W})\right] - 2\tilde{\varphi}\left(\pi''([\bar{Z},\varphi(\bar{W})])\right) + 2\tilde{\varphi}\left(\pi''([\bar{W},\varphi(\bar{Z})])\right).
\end{aligned}$$

Here $\pi'' : T(U) \otimes \mathbb{C} \longrightarrow T^{0,1}(U)$ is the projection with kernel $T^{1,0}(U)$. Recall that $Q : T(M) \otimes \mathbb{C} \longrightarrow T^{0,1}(M)$ is the projection with kernel E. We claim

$$\pi''|T(M) \otimes \mathbb{C} = Q - \tfrac{i}{2}\theta \otimes \bar{\xi}.$$

Indeed $\pi''(T) = \pi''\left(\frac{i}{2}(\xi - \bar{\xi})\right) = -\frac{i}{2}\bar{\xi}$. We obtain

$$\begin{aligned}
[\varphi,\varphi](\bar{Z},\bar{W}) &= 2\left[\varphi(\bar{Z}),\varphi(\bar{W})\right] - 2\varphi\left(Q([\bar{Z},\varphi(\bar{W})])\right) \\
&\quad +2\varphi\left(Q([\bar{W},\varphi(\bar{Z})])\right) - \tfrac{i}{2}\theta\left([\bar{Z},\varphi(\bar{W})] - [\bar{W},\varphi(\bar{Z})]\right)\tilde{\varphi}(\bar{\xi}).
\end{aligned}$$

But

$$\theta\left([\bar{Z},\varphi(\bar{W})] - [\bar{W},\varphi(\bar{Z})]\right) = \iota_\varphi d\theta(\bar{Z},\bar{W}) = 0.$$

Thus
$$R_2(\varphi)(\bar{Z},\bar{W}) = [\varphi,\varphi](\bar{Z},\bar{W}).$$
Since $[\varphi(\bar{Z}),\varphi(\bar{W})] \in \Gamma\left(M,T^{1,0}(M)\right)$ we have $R_3(\varphi)(\bar{Z},\bar{W}) = 0$ and the lemma follows. □

In the following the operator $\mathcal{L} : \mathcal{K}^q \longrightarrow \mathcal{K}^q$, $q \geq 1$, is defined by $\mathcal{L}\varphi = \varphi - \bar{\partial}_b\theta_\varphi$ where θ_φ is defined in Proposition 8.2. We use $D : \mathcal{B}^1 \longrightarrow \mathcal{B}^2$ to denote the restriction $\bar{\partial}_b|\mathcal{B}^1$. We choose a Hermitian metric on M and obtain adjoints $\bar{\partial}_b^* : \mathcal{K}^q \longrightarrow \mathcal{K}^{q-1}$ and $D^* : \mathcal{B}^2 \longrightarrow \mathcal{B}^1$. We let $H^q(\mathcal{K})$ denote the harmonic q-forms and H_b be the harmonic projection.

We now examine Miyajima's construction of the versal family for Kuranishi's CR-deformation theory. We first need a complement C^1 to B^1 in $\mathcal{B}^1$. We define $\mathcal{H} = \mathcal{L}H^1(\mathcal{K}^\cdot)$ hence
$$\mathcal{H} = \left\{\varphi - \bar{\partial}_b\theta_\varphi : \varphi \in H^1(\mathcal{K}^\cdot)\right\}.$$
We then define $C^1 \subset \mathcal{B}^1$ by
$$C^1 = \mathcal{H} + \operatorname{Im} D^*.$$

We observe that the sum is direct. Indeed suppose there exists $\varphi \in \mathcal{H}$ and $\psi \in \mathcal{B}^2$ such that $\varphi = D^*\psi$. By the Hodge decomposition for $\mathcal{B}^2$ we may assume ψ is exact. But φ is closed so $DD^*\psi = \Box\psi = 0$ so $\psi = 0$ (here $\Box = DD^* + D^*D$).

We now prove that C^1 is a complement to B^1 in $\mathcal{B}^1$. In what follows $N : \mathcal{B}^2 \longrightarrow \mathcal{B}^2$ denotes the Neumann operator, see [Mi1].

Lemma 9.6. *C^1 is a complement to B^1 in $\mathcal{B}^1$.*

Proof. We first observe that $\mathcal{H} \cap B^1 = \{0\}$. Indeed suppose $\varphi \in H^1(\mathcal{K}^\cdot)$ and $\varphi - \bar{\partial}_b\theta_\varphi \in B^1$. Then $\varphi \in B^1(\mathcal{K}^\cdot)$ whence $\varphi = 0$. □

We next claim that $B^1 \cap (\mathcal{H} + \operatorname{Im} D^*) = \{0\}$. Indeed suppose there exists $\lambda \in \mathcal{B}^0$, $\varphi \in \mathcal{H}$ and $\psi \in \mathcal{B}^2$ such that
$$\bar{\partial}_b\lambda = \varphi + D^*\psi.$$
Now we may assume ψ is orthogonal to $\ker D^*$, hence that ψ is exact (by the Hodge decomposition). We now apply D to the above equation to obtain $DD^*\psi = 0$ hence $\Box\psi = 0$. But ψ is exact hence $\psi = 0$.

We next prove that $\mathcal{B}^1 = B^1 + C^1$. Let $\varphi \in \mathcal{B}^1$. We claim there exists $\psi \in \mathcal{B}^2$ such that $\varphi - D^*\psi$ is closed. Indeed put $\psi = ND\varphi$. Then
$$D(\varphi - D^*ND\varphi) = D\varphi - DD^*ND\varphi.$$
But $DD^*N\omega = \omega$ for any exact $\omega \in \mathcal{B}^2$ and the claim follows. Finally since $\mathcal{H}$ contains a set of representatives for $H^1(\mathcal{B}^\cdot)$ we may choose $h \in \mathcal{H}$ such that $\varphi - D^*\psi - h \in B^1$. □

We now define $F : \mathcal{B}^1 \longrightarrow \mathcal{B}^1$, the Kuranishi map, by
$$F(\xi) = \xi + \tfrac{1}{2}D^*N\left([\xi,\xi]\right).$$

If $\hat{\mathcal{B}}^{\cdot}$ is a suitable Folland-Stein completion of $\mathcal{B}^{+}$ there is an induced map $\hat{F}$: $\hat{\mathcal{B}}^1 \longrightarrow \hat{\mathcal{B}}^1$ (this follows from the formula for $[\ ,\ ]$ and the fact that ξ takes values in $T^{1,0}(M)$. By the inverse function theorem $\hat{F}$ is invertible near zero. Hence there exist open subsets B and B' of $\hat{\mathcal{B}}$ containing the origin such that $\hat{F}(B) = B'$ and $\hat{F}$ is a biholomorphism from B to B'. Following [Mil], page 167 we define, for $B'' \subset B'$ a small ball around 0 in $\mathcal{B}^1$,

$$T(B'') = \left\{\eta \in B'' \cap \mathcal{H} : H_b\left([\hat{F}^{-1}(\eta), \hat{F}^{-1}(\eta)]\right) = 0\right\}.$$

We let $(T, 0)$ be the germ obtained by taking the limit with respect to B''. The main theorem of [Mil] is then the following (we assume $\dim M \geq 7$).

Theorem 9.3. *The analytic germ $(T, 0)$ is the parameter space of the versal deformation of M for Kuranishi's CR-deformation theory.*

Remark. We will use the symbol $\mathcal{K}_M$ instead of T henceforth and call $\mathcal{K}_M$ the Kuranishi space of M. The existence of an analytic structure on $\mathcal{K}_M$ can be proved following the usual "Kuranishi technique", [GM2], §2. We need to verify that $\mathcal{H} + D^*\hat{B}^2$ is closed. By a standard theorem in analysis $D^*\hat{B}^2$ is closed if and only if $D\hat{B}^1$ is closed. But $D\hat{B}^1$ has finite codimension in the closed subspace $\hat{Z}^2$ (the cocycles in $\hat{B}^2$) and consequently is closed in $\hat{Z}^2$ since it is the image of the bounded operator $D : \hat{B}^1 \longrightarrow \hat{Z}^2$ (see [GM3], page 497).

We can now take the last step in the proof of our main theorem.

Theorem 9.4. *The Kuranishi map $F : \mathcal{B}^1 \longrightarrow \mathcal{B}^1$ induces an isomorphism of functors $F : Y_{\hat{\mathcal{K}}} \longrightarrow \mathcal{K}_M$.*

The proof of Theorem 9.4 is essentially the same as that of Theorem 1.4. However we are forced to deal with the necessity to complete L^1 in order to invert F. This problem disappears on the infinitesimal level as we now check. Let $A \in \mathcal{A}$ with maximal ideal $\mathfrak{m}$. Recall that $F_A : L^1 \otimes \mathfrak{m} \longrightarrow L^1 \otimes \mathfrak{m}$ is the map induced by the polynomial mapping F. By the infinitesimal inverse function theorem, see [GM2], Lemma 3.1, F_A is one-to-one and onto hence invertible. We now describe the A-points of the analytic set $\mathcal{K}_M$.

Lemma 9.7. $\mathcal{K}_M(A) = \{\eta \in \mathcal{H} \otimes \mathfrak{m} : H_b([F_A^{-1}(\eta), F_A^{-1}(\eta)]) = 0\}$.

Proof. We have a commutative diagram

$$\begin{array}{ccc}
\hat{L}^1 \otimes \mathfrak{m} & \xrightarrow{\hat{F}_A} & \hat{L}^1 \otimes \mathfrak{m} \\
\uparrow & & \uparrow \\
L^1 \otimes \mathfrak{m} & \xrightarrow{F_A} & L^1 \otimes \mathfrak{m}
\end{array}$$

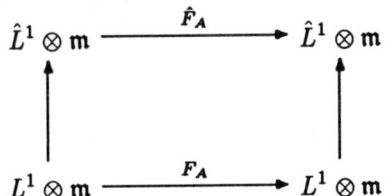

whence a commutative diagram

$$\begin{array}{ccc} \hat{L}^1 \otimes \mathfrak{m} & \xrightarrow{(\hat{F}_A)^{-1}} & \hat{L}^1 \otimes \mathfrak{m} \\ \uparrow & & \uparrow \\ L^1 \otimes \mathfrak{m} & \xrightarrow{F_A^{-1}} & L^1 \otimes \mathfrak{m} \end{array}$$

Thus if $\eta \in L^1 \otimes \mathfrak{m}$ we have

$$(\hat{F}_A)^{-1}(\eta) = F_A^{-1}(\eta).$$

Now since $\hat{F}^{-1}$ is analytic in a neighbourhood of zero it induces a formal map (see Chapter 1)

$$(\hat{F}^{-1})_A : \hat{L}^1 \otimes \mathfrak{m} \longrightarrow \hat{L}^1 \otimes \mathfrak{m}.$$

By definition

$$\mathcal{K}_M(A) = \left\{\eta \in \mathcal{H} \otimes \mathfrak{m} : H_b\left([(\hat{F}^{-1})_A(\eta), (\hat{F}^{-1})_A(\eta)]\right) = 0\right\}.$$

But we have seen that if F and G are composable formal maps then

$$(F \circ G)_A = F_A \circ G_A$$

whence

$$(\hat{F}^{-1})_A = (\hat{F}_A)^{-1}.$$

Hence if $\eta \in \mathcal{H} \otimes \mathfrak{m}$ we have

$$(\hat{F}^{-1})_A(\eta) = F_A^{-1}(\eta).$$

□

The proof of Theorem 9.4 is now the same as that of Theorem 1.4 using $\delta = D^* N$ and the splittings

$$\mathcal{B}^1 = B^1 + \mathcal{H} + \operatorname{Im} D^*$$

and

$$\mathcal{B}^2 = B^2 + H^2(\mathcal{B}) + \operatorname{Im}\left((\bar{\partial}_b | \mathcal{B}^2)^*\right).$$

We observe that $D^* N$ annihilates $\operatorname{Im}\left((\bar{\partial}_b|\mathcal{B}^2)^*\right)$ and $H^2(\mathcal{B})$ and assigns to $\beta \in B^2$ its unique coexact primitive. Thus $D^* N = \delta$ where δ is as in Chapter 1 and the definition of F_A here agrees with that of Chapter 1. Then Theorem 9.4 follows from Theorem 1.4. We have obtained the last result we need for our main theorem.

THEOREM F. *The differential graded Lie algebra $\bar{\mathcal{K}}$ is a controlling differential graded Lie algebra for Kuranishi's CR-deformation theory.* □

Remark. In the case $\operatorname{depth}_{\{0\}} V = 2$ the reader will check that the various homomorphisms of controlling differential graded Lie algebras induce isomorphisms on first cohomology. Hence we obtain an embedding of the base of the versal deformation of $(V, 0)$ into $\mathcal{K}_M$.

10. Counterexamples

In this chapter we show that the hypothesis $\mathrm{depth}_{\{0\}}\mathrm{V} \geq 3$ is necessary in order that the conclusion of the Main Theorem holds. The results in this chapter were obtained with the help of H. Flenner. Let $(Y, 0)$ be an irreducible analytic subset Y of a ball B around 0. We assume Y has 0 as its only singular point. We assume further that

(i) $\dim Y = n \geq 3$
(ii) $\mathrm{depth}_{\{0\}}\mathrm{Y} = 2$.

Now choose $f \in \mathcal{O}(Y)$ such that $f(0) = 0$ and $dy(y) \neq 0$, for all $y \in Z(f) - \{0\}$ where $Z(f) = \{y \in Y : f(y) = 0\}$. Put $X = Z(f)$. Then X has an isolated singularity at 0 and $\mathrm{depth}_{\{0\}}\mathrm{X} = 1$. By shrinking Y we may assume that $f : Y - \{0\} \longrightarrow \mathbb{C}$ is a submersion. Since f is not a zero divisor the map $f : Y \longrightarrow \mathbb{C}$ is flat and is a deformation of X. Let $n : \widetilde{T} \longrightarrow X$ be the normalization of X. Then n is finite and induces a biholomorphism $n : \widetilde{T} - n^{-1}(0) \longrightarrow X - \{0\}$.

As an example of the above construction we can take Y to be the cone on a projectively normal abelian variety of dimension 3 or greater and f to be a generic linear function on $\mathbb{C}^N$.

Our goal is to choose f such that the corresponding (see below) first order deformation of X cannot be lifted to a first order deformation of $\widetilde{T}$. Choose a point $\tilde{0} \in n^{-1}(0)$, let $A = \mathcal{O}_{X,0}$, $B = \mathcal{O}_{\widetilde{T},\tilde{0}}$ and $C = \mathcal{O}_{Y,0}$. The normalization map induces an embedding $n : A \longrightarrow B$. Now let $\tilde{A}$ be the deformation of $\mathcal{O}_{X,x}$ given by $\tilde{A} = C/(f^2)$. We obtain an exact sequence

$$0 \longrightarrow A\varepsilon \longrightarrow \tilde{A} \xrightarrow{\pi} A \longrightarrow 0$$

where π is reduction modulo (f) and $\varepsilon^2 = 0$. Such an extension corresponds to a deformation of A over $\mathbb{C}[\varepsilon]$, see [Sc3], and hence to an element $\tau \in T^1(A)$. We let $\mathfrak{m}$ and $\mathfrak{n}$ be the maximal ideals of A and B respectively. We obtain a long exact sequence of local cohomology

$$0 \longrightarrow H^1_{\mathfrak{m}}(A\varepsilon) \longrightarrow H^1_{\mathfrak{m}}(\tilde{A}) \longrightarrow H^1_{\mathfrak{m}}(A) \xrightarrow{\delta} H^2_{\mathfrak{m}}(A\varepsilon) \longrightarrow \cdots$$

We say that τ lifts to B if there exits an analytic local $\mathbb{C}$-algebra $\tilde{B}$ and a map $\tilde{n} : \tilde{A} \longrightarrow \tilde{B}$ and a commutative diagram with exact rows

$$\begin{array}{ccccccccc} 0 & \longrightarrow & A\varepsilon & \longrightarrow & \tilde{A} & \longrightarrow & A & \longrightarrow & 0 \\ & & \downarrow{\scriptstyle n} & & \downarrow{\scriptstyle \tilde{n}} & & \downarrow{\scriptstyle n} & & \\ 0 & \longrightarrow & B\varepsilon & \longrightarrow & \tilde{B} & \longrightarrow & B & \longrightarrow & 0 \end{array}$$

Then τ lifts to B means that we have a diagram of deformations of analytic local rings over $\mathbb{C}[\varepsilon]$

$$\begin{array}{ccc} \tilde{B} & \longrightarrow & B \\ \uparrow & & \uparrow \\ \tilde{A} & \longrightarrow & A \\ \uparrow & & \uparrow \\ \mathbb{C}[\varepsilon] & \longrightarrow & \mathbb{C} \end{array}$$

Lemma 10.1. *If τ lifts to B then the map $\delta : H^1_{\mathfrak{m}}(A) \longrightarrow H^2_{\mathfrak{m}}(A\varepsilon)$ is zero.*

Proof. The previous diagram of short exact sequences induces a diagram

$$\begin{array}{ccc} H^1_{\mathfrak{m}}(A) & \longrightarrow & H^2_{\mathfrak{m}}(A\varepsilon) \\ \downarrow & & \downarrow \\ H^1_{\mathfrak{n}}(B) & \longrightarrow & H^2_{\mathfrak{n}}(B\varepsilon) \end{array}$$

But $H^1_{\mathfrak{n}}(B) = 0$ since $\text{depth}_{\{0\}}\text{B} = 2$. Also the map $H^2_{\mathfrak{m}}(A\varepsilon) \longrightarrow H^2_{\mathfrak{n}}(B\varepsilon)$ is an isomorphism since n induces an isomorphism $\operatorname{Spec} A - \mathfrak{m} \longrightarrow \operatorname{Spec} B - \mathfrak{n}$. □

Thus our goal is accomplished if we can choose f such that δ above is non-zero. Now we observe that the exact sequence $C \longrightarrow C \longrightarrow C/(f^n)$ induced by multiplication by f^n induces an isomorphism $\delta_n : H^1_{\mathfrak{m}}(C/(f^n)) \xrightarrow{\simeq} \operatorname{Ann}_{f^n} H^2_{\bar{\mathfrak{m}}}(C)$ where $\bar{\mathfrak{m}}$ is the maximal ideal of C and $\operatorname{Ann}_g H^2_{\bar{\mathfrak{m}}}(C)$ means the annihilator of $g \in \bar{\mathfrak{m}}$ in $H^1_{\bar{\mathfrak{m}}}(C)$. Taking $n = 1$ or $n = 2$ respectively, we obtain a commutative diagram

$$\begin{array}{ccc} H^1_{\mathfrak{m}}(C/(f^2)) & \xrightarrow[\simeq]{\delta_2} & \operatorname{Ann}_{f^2} H^2_{\bar{\mathfrak{m}}}(C) \\ \rho \downarrow & & \downarrow \\ H^1_{\mathfrak{m}}(C/(f)) & \xrightarrow[\simeq]{\delta_1} & \operatorname{Ann}_f H^2_{\bar{\mathfrak{m}}}(C) \end{array}$$

where the right hand vertical arrow is multiplication by f and the left vertical arrow is reduction modulo (f).

Lemma 10.2. *Suppose the C-module $M_f = \operatorname{Ann}_f H^2_{\bar{\mathfrak{m}}}(C)/f \operatorname{Ann}_{f^2} H^2_{\bar{\mathfrak{m}}}(C)$ is non-zero. Then the infinitesimal deformation τ does not lift to B.*

Proof. As $\tilde{A} = C/(f^2)$ and $A = C/(f)$, the above module is the cokernel of the reduction map $\rho : H^1_{\mathfrak{m}}(\tilde{A}) \longrightarrow H^1_{\mathfrak{m}}(A)$. Hence δ induces an injection of M_f into $H^2_{\mathfrak{m}}(A)$. □

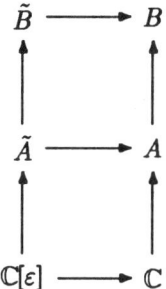

We observe that if Y is the cone on a projectively normal abelian variety N then C and $H^2_{\bar{\mathfrak{m}}}(C)$ are graded and we have

$$H^2_{\bar{\mathfrak{m}}}(C) = \bigoplus_{\nu \in \mathbb{Z}} H^1(N, \mathcal{O}(\nu)) = H^1(N, \mathcal{O}).$$

This last equality follows by the Kodaira Vanishing Theorem and Serre duality. Thus $H^2_{\bar{\mathfrak{m}}}(C)$ is concentrated in degree 0. Since multiplication by $f \in \bar{\mathfrak{m}}$ increases weight we find that $\bar{\mathfrak{m}}$ acts by zero on $H^2_{\bar{\mathfrak{m}}}(C)$. Hence for $f \in \bar{\mathfrak{m}}$, $\mathrm{Ann}_f\, H^1_{\bar{\mathfrak{m}}}(C) = H^2_{\bar{\mathfrak{m}}}(C)$ and $f\, \mathrm{Ann}_{f^2}\, H^2_{\bar{\mathfrak{m}}}(C) = \{0\}$. We have obtained the following lemma.

Lemma 10.3. *Let Y be the cone on a projectively normal abelian variety dimension 3 or greater. Let f be a generic linear function on Y. Then the infinitesimal deformation τ of $\mathcal{O}_{Y,y}/(f)$ corresponding to the extension*

$$0 \longrightarrow (f)/(f^2) \longrightarrow \mathcal{O}_{Y,y}/(f^2) \longrightarrow \mathcal{O}_{Y,y}/(f) \longrightarrow 0$$

does not lift to the normalization of $\mathcal{O}_{Y,y}/(f)$.

Let U be the complement of $n^{-1}(0)$ in X. Then there is a remarkable map $\mathrm{Def}(\tilde{X}, D)$ to $\mathrm{Def}(U, D)$ for any Artin local ring D, [Sc2]. Indeed a deformation $\tilde{X}_D$ of $\tilde{X}$ over $\mathrm{Spec}\, D$ is the space $\tilde{X}$ and a sheaf $\mathcal{O}_{\tilde{X}_D}$ which reduces to $\mathcal{O}_{\tilde{X}}$. The map above is then obtained by restricting the sheaf $\mathcal{O}_{\tilde{X}_D}$ to the open subset $U \subset \tilde{X}$. In case $D = \mathbb{C}[\varepsilon]$ we obtain the map $T^1(\tilde{X}) \longrightarrow H^1(U, \Theta)$ of [Sc2], page 152. We say a deformation $\tau \in \mathrm{Def}(U; D)$ extends over $\tilde{X}$ if it is in the image of previous map. Now observe that the deformation τ of A induces a deformation of a Stein representative X and thus by the above induces a deformation of $U = X - \{0\} = \tilde{X} - n^{-1}(0)$.

Lemma 10.4. *The deformation τ does not extend over $\tilde{X}$.*

Proof. Suppose the contrary. Then there exists a deformation

$$\begin{array}{ccc} \tilde{X} & \longrightarrow & \tilde{X}_{\mathbb{C}[\varepsilon]} \\ \downarrow & & \downarrow \\ \mathrm{Spec}\, \mathbb{C} & \longrightarrow & \mathrm{Spec}\, \mathbb{C}[\varepsilon] \end{array}$$

Since $\tilde{X}_{\mathbb{C}[\varepsilon]}$ is normal the identity map from $U_{\mathbb{C}[\varepsilon]}$ to itself extends to a map $\gamma : \tilde{X}_{\mathbb{C}[\varepsilon]}$ (see Lemma 9.1 of [Ar2]). But the restriction of γ to $\tilde{X}$ coincides with n. This contradicts Lemma 10.3. ☐

Since τ does not extend over $\tilde{X}$ there exists $x_i \in n^{-1}(0)$ such that τ does not extend over x_i. By the above we may take $x_i = \tilde{0}$. Choose a neighbourhood V of x_i in $\tilde{X}$ such that $V \cap n^{-1}(0) = \{x_i\}$. Then $U_{\mathbb{C}[\varepsilon]}$ induces a first-order deformation of $V - \{x_i\}$ that does not extend over V. We obtain the following theorem.

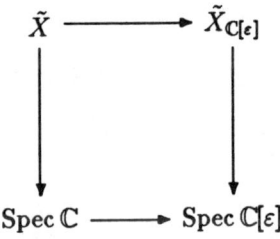

Theorem 10.1. *Let n be an integer satisfying $n \geq 3$. Then there exists an analytic subset V^n of a ball B around the origin in $\mathbb{C}^N$ such that 0 is the only singularity of V, $(V, 0)$ is normal and the natural map $T^1(V) \longrightarrow H^1(U, \Theta)$ is not onto where $U = V - \{0\}$. Moreover we may choose non-zero elements in the cokernel that are integrable.* □

Corollary. *Let M be a link of V. Then the natural map $T^1(V) \longrightarrow H^1(M, E)$ is not onto.*

□

REFERENCES

[Ak1] T. Akahori, *Intrinsic formula for Kuranishi's $\bar{\partial}_b$*, Publ. RIMS, Kyoto University **14** (1978), 615–641.

[Ak2] T. Akahori, *Complex analytic construction of the Kuranishi family on a normal strongly pseudo convex manifold*, Publ. RIMS, Kyoto University **14** (1978), 789–847.

[Ak3] T. Akahori, *The new estimate for the sub-bundles E_j and its application to the deformation of the boundaries of strongly pseudo-convex domains*, Invent. Math. **63** (1981), 311–344.

[Ak4] T. Akahori, *The new Neumann operator associated with deformation of strongly pseudo-convex domains and its application to deformation theory*, Invent. Math. **68** (1982), 317–352.

[AM] T. Akahori and K. Miyajima, *Complex analytic construction of the Kuranishi family on a normal strongly pseudo-convex manifold II*, Publ. RIMS, Kyoto University **18** (1980), 811–834.

[Ar1] M. Artin, *On solutions to analytic equations*, Invent. Math. **5** (1968), 277–291.

[Ar2] M. Artin, Lectures on Deformations of Singularities, Lectures on Mathematics and Physics **54**, Tata Institute (1976).

[Be] A. Beauville, *Foncteurs sur les anneaux Artiniens et applications aux deformations verselles*, Astérisque **16** (1974), 82–104.

[Bi] J. Bingener, Lokale Modulräume in der analytischen Geometrie, Aspekte der Mathematik, Band D2, Vieweg (1987).

[Bo1] N. Bourbaki, Commutative Algebra, Hermann (1972).

[Bo2] N. Bourbaki, Algebra II, Springer-Verlag (1980).

[D] P. Deligne, Letter to J.J. Millson, April 24, 1986.

[Fu] A. Fujiki, *Flat Stein completion of a flat $(1,1)$-convex concave map*, preprint.

[F] H. Flenner, *Über Deformationen holomorpher Abildungen*, Mathematisches Institut der Universität Göttingen, Nachdruck 1988.

[Fi] G. Fischer, *Complex Analytic Geometry*, Lecture Notes in Math **538**, Springer, New York, 1976.

[FN1] A. Frölicher and A. Nijenhuis, *Theory of vector-valued forms, Part I. Derivations in the graded ring of differential forms*, Proc. Kon. Ned. Akad. Wetensch. **A59** (1956), 338–359.

[FN2] A. Frölicher and A. Nijenhuis, *Some new cohomology invariants for complex manifolds*, Proc. Kon. Ned. Akad. Wetensch. **A59** (1956), 540–564.

[Go] R. Godement, Topologie algébrique et Théorie des Faisceaux, Hermann, Paris 1964.

[Gr1] H. Grauert, *Über die Deformationen isolierter Singularitäten analytischer Mengen.*, Invent. Math. **15** (1972), 171–198.

[Gr2] H. Grauert, *Der Satz von Kuranishi für kompakte komplexe Räume*, Invent. Math. **25** (1974), 107–142.

[Gr Mo] P. Griffiths and J. Morgan, Rational Homotopy Theory and Differential Forms, Progress in Mathematics **16**, Birkhäuser, Boston 1981.

[GM1] W.M. Goldman and J.J. Millson, *The deformation theory of representations of fundamental groups of compact Kähler manifolds*, Publ. Math. IHES **67** (1988), 43–96.

[GM2] W.M. Goldman and J.J. Millson, *The homotopy invariance of the Kuranishi space, III.* J. Math. **34** (1990), 337–367.

[GM3] W.M. Goldman and J.J. Millson, *Differential graded Lie algebras and singularities of level sets of momentum mappings*, Commun. Math. Phys. **131** (1990), 495–515.

[I] L. Illusie, *Complexe cotangent et Déformations I*, Lecture Notes in Math. **239**, Springer, New York, 1970.

[K1] M. Kuranishi, Deformations of Compact Complex Manifolds, Les Presses de l'Université de Montréal (1971).

[K2] M. Kuranishi, *Application of $\bar{\partial}_b$ to deformation of isolated singularities*, Proceedings of Symposia in Pure Mathematics **30** (1977), 97–106.

[Ma] B. Malgrange, Ideals of Differentiable Functions, Oxford University Press (1966).

[Mac] S. MacLane, Homology, Grundlehren der mathematischen Wissenschaften, Band **114**, Springer, New York, 1967.

[M1] J. Millson, *Rational homotopy theory and deformation problems from algebraic geometry*, Proc. ICM Kyoto 1990.

[M2] J. Millson, *CR-geometry and deformations of cones*, in preparation.

[M3] J. Millson, *CR-geometry and deformations of isolated singularities*, Proceedings of Symposia in Pure Math. **54** (1993), part 2, 411–433.

[Mi1] K. Miyajima, *Completion of Akahori's construction of the versal family of strongly pseudo-convex CR structures*, Trans. A.M.S. **277** (1983), 163–172.

[Mi2] K. Miyajima, *Deformation of a complex manifold near a strongly pseudo-convex real hypersurface and a realization of the Kuranishi family of strongly pseudo-convex CR-structures*, Math. Z. **205** (1990), 593–602.

[P] V. Palamodov, *Deformations of complex spaces*, Russian Math. Surveys **31:3** (1976), 129–197.

[S] M. Schneider, *Vollständige Durchschnitte in Steinschen Mannigfaltigkeiten*, Math. Ann. **186** (1970), 191–200.

[Sa] D.J. Saunders, *The Geometry of Jet Bundles*, London Math. Soc. Lecture Note Series **145**, Cambridge University Press, 1989.

[Sc1] M. Schlessinger, *Functors of Artin rings*, Trans. A.M.S. **130** (1966), 208–222.

[Sc2] M. Schlessinger, *On rigid singularities*, Conference on Complex Analysis, Rice University Studies **59**, No. 1 (1972), 147–162.

[Sc3] M. Schlessinger, *Infinitesimal deformation of singularities*, Ph.D. thesis, Harvard University.

[Schu] H. Schuster, *Infinitesimale Erweiterungen Komplexer Räume*, Comm. Math. Helv. **45** (1970), 265–286.

[SS1] M. Schlessinger and J. Stasheff, *Deformation theory and rational homotopy type*, preprint.

[SS2] M. Schlessinger and J. Stasheff, *The Lie algebra structure of tangent cohomology and deformation theory*, Journal of Pure and Appl. Algebra **38** (1985), 313–322.

[Ta] N. Tanaka, A Differential-Geometric Study on Strongly Pseudo-Convex Manifolds, Lectures in Mathematics **9**, Kyoto University, 1975.

[Y] Stephen S.-T, Yau, *Kohn-Rossi cohomology and its application to the complex Plateau problem I*, Ann. of Math. **113** (1981), 67–110.

Ragnar-Olaf Buchweitz
DEPARTMENT OF MATHEMATICS, UNIVERSITY OF TORONTO, TORONTO, ONTARIO, M5S 1A7, CANADA
E-mail address: ragnar@math.utoronto.ca

John J. Millson
DEPARTMENT OF MATHEMATICS, UNIVERSITY OF MARYLAND, COLLEGE PARK, MD 20742, USA
E-mail address: jjm@math.umd.edu

Editorial Information

To be published in the *Memoirs*, a paper must be correct, new, nontrivial, and significant. Further, it must be well written and of interest to a substantial number of mathematicians. Piecemeal results, such as an inconclusive step toward an unproved major theorem or a minor variation on a known result, are in general not acceptable for publication. *Transactions* Editors shall solicit and encourage publication of worthy papers. Papers appearing in *Memoirs* are generally longer than those appearing in *Transactions* with which it shares an editorial committee.

As of September 30, 1996, the backlog for this journal was approximately 7 volumes. This estimate is the result of dividing the number of manuscripts for this journal in the Providence office that have not yet gone to the printer on the above date by the average number of monographs per volume over the previous twelve months, reduced by the number of issues published in four months (the time necessary for preparing an issue for the printer). (There are 6 volumes per year, each containing at least 4 numbers.)

A Copyright Transfer Agreement is required before a paper will be published in this journal. By submitting a paper to this journal, authors certify that the manuscript has not been submitted to nor is it under consideration for publication by another journal, conference proceedings, or similar publication.

Information for Authors and Editors

Memoirs are printed by photo-offset from camera copy fully prepared by the author. This means that the finished book will look exactly like the copy submitted.

The paper must contain a *descriptive title* and an *abstract* that summarizes the article in language suitable for workers in the general field (algebra, analysis, etc.). The *descriptive title* should be short, but informative; useless or vague phrases such as "some remarks about" or "concerning" should be avoided. The *abstract* should be at least one complete sentence, and at most 300 words. Included with the footnotes to the paper, there should be the 1991 *Mathematics Subject Classification* representing the primary and secondary subjects of the article. This may be followed by a list of *key words and phrases* describing the subject matter of the article and taken from it. A list of the numbers may be found in the annual index of *Mathematical Reviews*, published with the December issue starting in 1990, as well as from the electronic service e-MATH [**telnet e-MATH.ams.org** (or **telnet 130.44.1.100**). Login and password are **e-math**]. For journal abbreviations used in bibliographies, see the list of serials in the latest *Mathematical Reviews* annual index. When the manuscript is submitted, authors should supply the editor with electronic addresses if available. These will be printed after the postal address at the end of each article.

Electronically prepared papers. The AMS encourages submission of electronically prepared papers in $\mathcal{A}\mathcal{M}\mathcal{S}$-TeX or $\mathcal{A}\mathcal{M}\mathcal{S}$-LaTeX. The Society has prepared author packages for each AMS publication. Author packages include instructions for preparing electronic papers, the *AMS Author Handbook*, samples, and a style file that generates the particular design specifications of that publication series for both $\mathcal{A}\mathcal{M}\mathcal{S}$-TeX and $\mathcal{A}\mathcal{M}\mathcal{S}$-LaTeX.

Authors with FTP access may retrieve an author package from the Society's Internet node `e-MATH.ams.org` (130.44.1.100). For those without FTP access, the author package can be obtained free of charge by sending e-mail to `pub@math.ams.org` (Internet) or from the Publication Division, American Mathematical Society, P.O. Box 6248, Providence, RI 02940-6248. When requesting an author package, please specify $\mathcal{A}\mathcal{M}\mathcal{S}$-TeX or $\mathcal{A}\mathcal{M}\mathcal{S}$-LaTeX, Macintosh or IBM (3.5) format, and the publication in which your paper will appear. Please be sure to include your complete mailing address.

Submission of electronic files. At the time of submission, the source file(s) should be sent to the Providence office (this includes any TeX source file, any graphics files, and the DVI or PostScript file).

Before sending the source file, be sure you have proofread your paper carefully. The files you send must be the EXACT files used to generate the proof copy that was accepted for publication. For all publications, authors are required to send a printed copy of their paper, which exactly matches the copy approved for publication, along with any graphics that will appear in the paper.

TeX files may be submitted by email, FTP, or on diskette. The DVI file(s) and PostScript files should be submitted only by FTP or on diskette unless they are encoded properly to submit through e-mail. (DVI files are binary and PostScript files tend to be very large.)

Files sent by electronic mail should be addressed to the Internet address `pub-submit@math.ams.org`. The subject line of the message should include the publication code to identify it as a Memoir. TeX source files, DVI files, and PostScript files can be transferred over the Internet by FTP to the Internet node `e-math.ams.org` (130.44.1.100).

Electronic graphics. Figures may be submitted to the AMS in an electronic format. The AMS recommends that graphics created electronically be saved in Encapsulated PostScript (EPS) format. This includes graphics originated via a graphics application as well as scanned photographs or other computer-generated images.

If the graphics package used does not support EPS output, the graphics file should be saved in one of the standard graphics formats—such as TIFF, PICT, GIF, etc.—rather than in an application-dependent format. Graphics files submitted in an application-dependent format are not likely to be used. No matter what method was used to produce the graphic, it is necessary to provide a paper copy to the AMS.

Authors using graphics packages for the creation of electronic art should also avoid the use of any lines thinner than 0.5 points in width. Many graphics packages allow the user to specify a "hairline" for a very thin line. Hairlines often look acceptable when proofed on a typical laser printer. However, when produced on a high-resolution laser imagesetter, hairlines become nearly invisible and will be lost entirely in the final printing process.

Screens should be set to values between 15% and 85%. Screens which fall outside of this range are too light or too dark to print correctly.

Any inquiries concerning a paper that has been accepted for publication should be sent directly to the Editorial Department, American Mathematical Society, P. O. Box 6248, Providence, RI 02940-6248.

Editors

Selected Titles in This Series

(Continued from the front of this publication)

566 **Neil Robertson, Paul Seymour, and Robin Thomas,** Excluding infinite clique minors, 1995
565 **Huaxin Lin and N. Christopher Phillips,** Classification of direct limits of even Cuntz-circle algebras, 1995
564 **Wensheng Liu and Héctor J. Sussmann,** Shortest paths for sub-Riemannian metrics on rank-two distributions, 1995
563 **Fritz Gesztesy and Roman Svirsky,** (m)KdV solitons on the background of quasi-periodic finite-gap solutions, 1995
562 **John Lindsay Orr,** Triangular algebras and ideals of nest algebras, 1995
561 **Jane Gilman,** Two-generator discrete subgroups of $PSL(2, R)$, 1995
560 **F. Tomi and A. J. Tromba,** The index theorem for minimal surfaces of higher genus, 1995
559 **Paul S. Muhly and Baruch Solel,** Hilbert modules over operator algebras, 1995
558 **R. Gordon, A. J. Power, and Ross Street,** Coherence for tricategories, 1995
557 **Kenji Matsuki,** Weyl groups and birational transformations among minimal models, 1995
556 **G. Nebe and W. Plesken,** Finite rational matrix groups, 1995
555 **Tomás Feder,** Stable networks and product graphs, 1995
554 **Mauro C. Beltrametti, Michael Schneider, and Andrew J. Sommese,** Some special properties of the adjunction theory for 3-folds in $\mathbb{P}^5$, 1995
553 **Carlos Andradas and Jesús M. Ruiz,** Algebraic and analytic geometry of fans, 1995
552 **C. Krattenthaler,** The major counting of nonintersecting lattice paths and generating functions for tableaux, 1995
551 **Christian Ballot,** Density of prime divisors of linear recurrences, 1995
550 **Huaxin Lin,** C^*-algebra extensions of $C(X)$, 1995
549 **Edwin Perkins,** On the martingale problem for interactive measure-valued branching diffusions, 1995
548 **I-Chiau Huang,** Pseudofunctors on modules with zero dimensional support, 1995
547 **Hongbing Su,** On the classification of C*-algebras of real rank zero: Inductive limits of matrix algebras over non-Hausdorff graphs, 1995
546 **Masakazu Nasu,** Textile systems for endomorphisms and automorphisms of the shift, 1995
545 **John L. Lewis and Margaret A. M. Murray,** The method of layer potentials for the heat equation on time-varying domains, 1995
544 **Hans-Otto Walther,** The 2-dimensional attractor of $x'(t) = -\mu x(t) + f(x(t-1))$, 1995
543 **J. P. C. Greenlees and J. P. May,** Generalized Tate cohomology, 1995
542 **Alouf Jirari,** Second-order Sturm-Liouville difference equations and orthogonal polynomials, 1995
541 **Peter Cholak,** Automorphisms of the lattice of recursively enumerable sets, 1995
540 **Vladimir Ya. Lin and Yehuda Pinchover,** Manifolds with group actions and elliptic operators, 1994
539 **Lynne M. Butler,** Subgroup lattices and symmetric functions, 1994
538 **P. D. T. A. Elliott,** On the correlation of multiplicative and the sum of additive arithmetic functions, 1994
537 **I. V. Evstigneev and P. E. Greenwood,** Markov fields over countable partially ordered sets: Extrema and splitting, 1994
536 **George A. Hagedorn,** Molecular propagation through electron energy level crossings, 1994

(See the AMS catalog for earlier titles)